W9-CNF-514

INVARIANT MEANS on Topological Groups and Their Applications

by

FREDERICK P. GREENLEAF
New York University

VAN NOSTRAND · REINHOLD COMPANY
NEW YORK TORONTO LONDON MELBOURNE

VAN NOSTRAND REGIONAL OFFICES: *New York, Chicago, San Francisco*

D. VAN NOSTRAND COMPANY, LTD., *London*

D. VAN NOSTRAND COMPANY (Canada), LTD., *Toronto*

D. VAN NOSTRAND AUSTRALIA PTY. LTD., *Melbourne*

Published simultaneously in Canada by
D. VAN NOSTRAND COMPANY (Canada), LTD.

PRINTED IN THE UNITED STATES OF AMERICA

PREFACE

Translation invariant Banach means on spaces of functions associated with a topological group have interested generations of mathematicians since the appearance of von Neumann's article [72], which deals with discrete groups. These invariant means are generally created by highly non-constructive methods—most often by invoking the Hahn-Banach Theorem—and have many strange properties, thus they are often thought of as mathematical curiosities. However, in recent years, some remarkably diverse properties of locally compact groups have been found to depend on the existence of a Banach invariant mean on an appropriate translation-invariant space of functions. One of the most striking results is the following.

Theorem: If G is a locally compact group, there exists a left invariant mean on $L^\infty(G)$ if and only if every irreducible unitary representation of G is weakly contained in the left regular representation.

We give a self-contained exposition, accessible to anyone with a modest understanding of functional analysis, of this and many other recent discoveries relating existence of invariant means to algebraic and geometric properties of a locally compact group.

In the past authors have considered invariant means on a number of spaces of functions; for example, Hulanicki [37] discusses left invariant means on L^∞ in studying the weak containment property above, Rickert [66] shows that the "fixed point property" for G is tied to the existence of a left invariant mean on the bounded right uniformly continuous functions on G (see section 3.3), and Glicksberg [21], Reiter [58] relate ergodic properties of G to existence of an invariant mean on the space $CB(G)$ of bounded continuous functions (see section 3.7). The connection between these various types of invariant means is not at all apparent. We shall prove (combining recent work of the author and Namioka) that these diverse notions of invariant mean are all equivalent for locally compact groups (Theorem 2.2.1); using this equivalence we shall unify many results in the literature and divest them of restrictive hypotheses.

In the past there have been several papers which recount the then current state of the literature: the articles by Dixmier [11], Day [8], and Hewitt-Ross [34] (sections 17-18) are quite helpful and are accessible to most mathematicians. All were written before the equivalence of invariant means was recognized (a very recent development) and before the most important applications had appeared in the literature; we present direct, self-contained accounts of these modern developments. Some of these results are difficult to extract from the scattered literature on invariant means, and many are presented with new proofs, simpler than those which appear in the literature.

These notes are based on lectures presented at Berkeley in the Spring and Fall 1966 quarters. They have benefitted greatly from the author's correspondence with A. Hulanicki, I. Namioka, E. Granirer and W. R. Emerson, and from numerous conversations with colleagues and visiting faculty at Berkeley.

Frederick P. Greenleaf

CONTENTS

Special Symbols

$\mathbf{R}$	real numbers
$\mathbf{C}$	complex numbers
$\mathbf{Z}$	integers
${}_xf(s) = f(x^{-1}s)$	[f defined on a group]
$L_xf(s) = f(xs)$	[f defined on any semigroup or group]
$R_xf(s) = f(sx)$	[f defined on any semigroup or group]
$\mathrm{Re} f$	real part of a function f
$A \backslash B$	difference of sets A, B
$A \Delta B$	$(A \backslash B) \cup (B \backslash A)$, the symmetric difference of A, B
$\|A\|$	measure (or cardinality) of a set A
χ_A	characteristic function of A
ϕ_A	$\frac{1}{\|A\|}\chi_A$, normalized characteristic function of A
δ_x	Dirac measure [point mass] at x

SECTION 1

INVARIANT MEANS ON DISCRETE GROUPS AND SEMIGROUPS

§1.1. MEANS AND INVARIANT MEANS

Let G be any set and X a closed subspace of $B(G)$, the space of *all* bounded complex-valued functions on G equipped with the sup norm $\|f\|_\infty$. Assume X includes all constant functions and is also closed under complex conjugation. Then a linear functional m on X is a *mean* if

(1) $\qquad m(\bar{f}) = \overline{m(f)} \qquad$ all $f \in X$.

(2) $\qquad \inf\{f(x)\} \le m(f) \le \sup\{f(x)\}$ for all real-valued $f \in X$.

The second condition is equivalent to

(2′) $\qquad m(f) \ge 0$ if $f \ge 0$, and $m(1) = 1$.

Thus (2) insures that $m(1) = 1$ and $\|m\| = 1$ for any mean. The means on X form a weak*-compact convex set Σ in X^*. If $\ell^1(G)$ is the space of all bounded discrete measures on G with total variation norm, then $B(G) = (\ell^1)^*$; obviously every non-negative measure $\mu \in \ell^1$ with $\|\mu\| = 1$ gives a mean on X: $m_\mu(f) = \langle \mu, f \rangle$ and these form a convex subset $\Sigma_d \subset \Sigma$, the set of *discrete means* on X. Furthermore, every mean m on S is the weak $*$ limit of some net of discrete means; otherwise the Hahn-Banach theorem (as in [13] V. 2.10) insures we can find a $\delta > 0$ and some $f \in X$ such that $\mathrm{Re}(m(f)) \ge \delta + \mathrm{Re}(m'(f))$ for all discrete means $m' \in \Sigma_d$. But since Σ_d includes all point masses, and $\mathrm{Re}(m(f)) = m(\mathrm{Re} f)$, we see that

$$m(\mathrm{Re} f) > \sup\{m'(\mathrm{Re} f)\colon m' \in \Sigma_d\} \geq \sup\{\mathrm{Re} f(x)\colon x \in G\}$$

which contradicts the definition of m being a mean. A similar argument applies to show density of the *finite means*: Σ_{fin}, those arising from measures which are finite sums of point masses.

If G is a group and if the function space X is left invariant, so $f \in X \Longrightarrow {}_xf \in X$, where ${}_xf(t) = f(x^{-1}t)$,[1] then a mean m is *left invariant* (m a LIM) if

(3) $\quad m({}_xf) = m(f)$ all $x \in G$, all $f \in X$.

Likewise we say m is a *right invariant* mean if $m(f_x) = m(f)$ for all $x \in G$, where we define $f_x(t) = f(tx)$, and we define two-sided invariance in the usual way, assuming of course X is invariant under right and left translations.

There is an interesting duality between right and left invariant means if G is a group. For $f \in X$ define $f^{\sim}(x) = f(x^{-1}$ In many cases of interest $X = X^{\sim}$; in any event we have:

Lemma 1.1.1. If G is a group, there is a left invariant mean on $X \Longleftrightarrow$ there is a right invariant mean on $X^{\sim}$.

Proof. Given left invariant mean m on X, define $\bar{m}$ on $X^{\sim}$ so $\bar{m}(f) = m(f^{\sim})$. It is easily verified that

$$(f^{\sim})_x = ({}_xf)^{\sim}$$

---

[1] If G is a semigroup there is a slightly different notion of left translation: $L_s f(t) = f(st)$. If X is left invariant, in the sense that $f \in X \Longrightarrow L_s f \in X$, and if G is actually a group, it is also left invariant in the above sense because $G = G^{-1}$. For functorial reasons we take the above definition of ${}_xf$ when G is a group: this way the action of G on X becomes a group representation instead of an anti-representation

$$\bar{m}((f^{\sim})_x) = \bar{m}(({}_xf)^{\sim}) = m({}_xf) = m(f) = \bar{m}(f^{\sim}) .$$

Example 1.1.2. For a semi-group, there is no analog of this result: let G be a non-empty set with product $xy = y$ for $x, y \in G$. Then if $f \in B(G)$ we have $L_x f(t) = f(xt) = f(t)$ so *every* mean on $B(G)$ is left-invariant. But $f_x(t) = f(tx) = f(x)\cdot 1$ so if m is a right invariant mean: $m(f) = m(f_x) = m(f(x)\cdot 1) = f(x)$ for all $x \in G$; if G has more than one element, this is clearly impossible.

In many cases existence of left and right invariant means m_ℓ and m_r insures existence of a two-sided invariant mean m. The general idea of the proof, which makes sense if $X = B(G)$ for example, is to take $f \in X$, define $F(x) = < m_\ell, f_x >$, and set $m(f) = m_r(F)$. It is readily verified that m is two-sided invariant, which proves:

Lemma 1.1.3. If G is a semigroup with a left invariant mean and a right invariant mean on $X = B(G)$, then there exists a two-sided invariant mean on X.

However, this construction does not always make sense, for if m_ℓ and m_r are invariant means on $X = CB(G)$: the continuous bounded functions on a topological group G, then there is no assurance that $f \in CB(G) \Longrightarrow F(x) = < m_\ell, f_x >$ is in $CB(G)$. On the other hand if X is a space of suitably uniformly continuous functions on G, there is some hope of making this construction work. A few constructions for two-sided invariant means are discussed in the introductory section of [36].

Our main interest is in left and right invariant means on *groups* and for applications it is only important to know whether there is at least one such invariant mean on X; the uniqueness of such means is not relevant. In view of the duality exhibited in 1.1.1, we shall generally discuss left invariant means when dealing with groups. It is fortunate that uniqueness of invariant means is not important in applications because they are usually not unique. If $X = B(G)$ and G is a finite group, or if X is a reasonable space of continuous functions and G is a compact group, then the normalized Haar measure on G gives a left invariant mean (LIM) on X and it is easily seen that this is the only LIM on X. The question of uniqueness has received a great deal of study; see Day [8], sections 6–7, and also Hewitt-Ross [34], section 17.21. Some recent results, especially those in Granirer [26], allow us to prove the following definitive result.

Theorem 1.1.5. Let G be any discrete group which admits a LIM on $B(G)$. Then $B(G)$ has a unique LIM $\Longleftrightarrow$ G is finite.

We prove this in Appendix 1. The situation is incompletely understood for invariant means on spaces of continuous functions on a non-discrete topological group, cf. [27]. In 2.4, once we have developed techniques for constructing invariant means we shall give some direct constructions of distinct invariant means (the methods of Appendix 1 are probabilistic).

§1.2 CONSTRUCTION OF INVARIANT MEANS

A discrete semigroup G is *left (right) amenable* if there is a left (right) invariant mean on $X = B(G)$; if G is a group these conditions are the same and we say that G is *amenable*. Our first problem is to find reasonable conditions on G which enable us (modulo the Hahn-Banach Theorem) to construct invariant means on $B(G)$. Dixmier [11] shows, following ideas which first appear in von Neumann [72], that existence of a LIM on $B(G)$ is equivalent to the following property of G.

(D) If $\{f_1, \ldots, f_N\}$ are real-valued functions in $B(G)$ and if $\{x_1, \ldots, x_N\} \subset G$ then:

$$\inf\{\sum_{i=1}^{N} (f_i - L_{x_i} f_i)\} \leq 0 .$$

Let X be the closed (real) subspace generated by $\{f - L_x f: x \in G, f$ real-valued$\}$. If m is a LIM on $B(G)$ it must annihilate X, so $0 = m(\phi) \geq \inf(\phi)$, all $\phi \in X$. Conversely if $\inf(\phi) \leq 0$ for all $\phi \in X$ then in $B_r(G)$, the real-valued bounded functions, consider $K = \{\phi \in B_r(G): \inf(\phi) > 0\}$. This open convex set is disjoint from the subspace X so by one form of Hahn-Banach there is a bounded linear functional m on $B_r(G)$ such that $m(X) = 0$, $m(f) > 0$ for all $f \in K$ By scaling we can arrange that $m(1) = 1$; thus m is a LIM on $B_r(G)$. Extend it to a LIM on $B(G)$ by taking $\bar{m}(f + ig) = m(f) + i \cdot m(g)$.

There are obvious right-handed and two-sided versions of this result, whose proofs we omit. Using this criterion we prove a basic existence theorem (following [34]; there is a gap in the proof which appears in [11]).

Theorem 1.2.1. There is an invariant measure on $B(G)$ for any abelian semigroup G.

Proof. Let $\{f_1, \ldots, f_N\} \subset B(G)$ and $\{x_1, \ldots, x_N\} \subset G$ be given. Write $\Lambda_p = \{(\lambda_1, \ldots, \lambda_N): \lambda_k$ integers, $1 \leq \lambda_k \leq p\}$ for $p = 1, 2, \ldots$, so that Λ_p has cardinality $|\Lambda_p| = p^N$, and define $\tau(\lambda) = x_1^{\lambda_1} \cdots x_N^{\lambda_N} \in G$. In any sum of the form

$$\Sigma\{f_k(\tau(\lambda)) - f_k(x_k \cdot \tau(\lambda)) : \lambda \in \Lambda_p\}$$

all terms cancel except possibly those $f_k(\tau(\lambda))$ with $\lambda_k = 1$ and those $f_k(x_k \cdot \tau(\lambda))$ with $\lambda_k = p$ (there are only $2p^{N-1}$ such λ in Λ_p for each k). But if $\phi(t) = \Sigma_{k=1}^N f_k(t) - f_k(x_k t)$ and $m = \max\{\|f_k\|_\infty : k = 1, 2, \ldots N\}$, then

$$|\Lambda_p| \inf\{\phi(t)\colon t \in G\} = p^N \inf\{\phi\} \le \Sigma\{\phi(\tau(\lambda))\colon \lambda \in \Lambda_p\}$$
$$= \sum_{k=1}^{N} \{\Sigma\{f_k(\tau(\lambda)) - f_k(x_k\tau(\lambda))\colon \lambda \in \Lambda_p\}\}$$
$$\le \sum_{k=1}^{N} 2p^{N-1}\|f_k\|_\infty \le 2mN\,p^{N-1} \qquad p = 1,2\ldots .$$

Thus $\inf\{\phi\} \le 0$ as required. Q.E.D.

If G is a finite group, then there is a (unique) LIM on $B(G)$ corresponding to Haar measure, but if G is a finite semigroup there may not be any LIM on $B(G)$ as 1.1.2 shows. Rosen [67] has characterized the finite semigroups which are two-sided amenable as follows—we will not prove this here.

Proposition 1.2.2. A finite semigroup G has a two-sided invariant mean on $B(G)$ $\Longleftrightarrow$ G has unique minimal left and right ideals; then these minimal ideals coincide in a two-sided ideal which is a finite group G^*. The (unique) invariant mean m on $B(G)$ is given by

$$m(f) = \frac{1}{|G^*|} \Sigma\{f(t)\colon t \in G^*\}$$

where $|G^*|$ = cardinality of G^*.

Example 1.2.3. If m is a left invariant mean on $B(G)$ we may define a left-invariant finitely additive measure μ on the collection $\Omega(G)$ of all subsets: $\mu(E) = m(\chi_E)$, where χ_E = characteristic function of $E \in \Omega(G)$. If G is the free group on two generators a, b such a measure cannot exist, thus G is not amenable: divide G into disjoint sets $\{H_i\colon i \in \mathbf{Z}\}$ with $x \in H_i$ $\Longleftrightarrow$ when expressed as a reduced word, $x = a^i b^{i_1} a^{i_2} \ldots$

($i_1 \neq 0$ if $x \neq a^i$). Then the transformation λ_a: $x \to ax$ maps H_i congruent to H_{i+1} for all $i \in \mathbf{Z}$ while λ_b: $x \to bx$ maps every set H_i ($i \neq 0$) into H_0. If a left invariant measure μ existed the first fact insures that $\mu(H_i) = 0$ all $i \in \mathbf{Z}$ (recall $\mu(G) = 1$); the second shows $\mu(H_0) \geq \mu(\cup_{i\neq 0} H_i)$ while $\mu(H_0) + \mu(\cup_{i\neq 0} H_i) = 1$, which $\Longrightarrow$ $\mu(H_0) \geq \frac{1}{2}$.

Now let G be a group (although much of what we say applies equally well to semi-groups). The study of invariant means on $B(G)$ was initiated by von Neumann [72] in his investigation of the Banach-Tarski paradox and related matters. He proved the following basic combinatory facts; we shall see that these results generalize to locally compact groups, but there are many analytic complications in adapting the simple arguments below.

Theorem 1.2.4. If G is amenable and π a homomorphism onto group H, then H is amenable.

Proof: Then $f \in B(H) \Longrightarrow f \circ \pi \in B(G)$ and $\overline{m}(f) = m(f \circ \pi)$ is evidently a LIM on $B(H)$. Q.E.D.

Theorem 1.2.5. Every subgroup H of an amenable group G is amenable.

Proof. If m is the LIM on $B(G)$, let $T = \{x_\alpha\colon \alpha \in A\}$ be a transversal for the right cosets $\{Hx\colon x \in G\}$, and for each $f \in B(H)$ consider the function $Tf \in B(G)$ gotten by transferring f to each right coset:

$$Tf(hx_\alpha) = f(h) \qquad \text{all } h \in H,\ \alpha \in A\,.$$

Define $\overline{m}(f) = m(Tf)$. It is trivial that $\overline{m}$ is a mean on $B(H)$ and that $T({}_hf) = {}_h(Tf)$ for all $h \in H$, which insures $\overline{m}$ is a LIM.

Q.E.D.

Theorem 1.2.6. If N is a normal subgroup in G and if N and G/N are amenable, then G is amenable.

Proof: Let m_1, m_2 be left invariant means on $B(N)$, $B(G/N)$. If $f \in B(G)$ then $xN = yN \Longrightarrow x = n'y$ for some $n' \in N$ and ${}_xf(n) = {}_{n'y}f(n)$ all $n \in N$ which $\Longrightarrow$ if we consider the functions ${}_xf$, ${}_yf$ restricted to N,

$$m_1({}_xf) = m_1({}_{n'}({}_yf)) = m_1({}_yf)$$

whenever x, y lie in the same coset of N. Thus $F(x) = m_1({}_xf)$ is constant on cosets of N and, when regarded as a function in $B(G/N)$, allows us to define

$$\overline{m}(f) = m_2(F) = < m_2(x), m_1({}_xf) >$$

as the desired LIM on $B(G)$.

Q.E.D.

Theorem 1.2.7. If G is a directed union of a system of amenable groups $\{H_\alpha\}$, in the sense that $G = \cup_\alpha H_\alpha$ and for any H_α, H_β there exists $H_\gamma \supset H_\alpha \cup H_\beta$, then G is amenable.

Proof. Let m_α be a LIM on $B(H_\alpha)$. Then $\overline{m}_\alpha(f) = m_\alpha(f|H_\alpha)$ gives a mean on $B(G)$ which is invariant under left translation by elements of H_α. Now the set Λ_α of means on $B(G)$ invariant under H_α is compact and the Λ_α have the finite intersection property, which $\Longrightarrow \Lambda = \cap_\alpha \Lambda_\alpha \neq \emptyset$. But every mean in Λ is a LIM on $B(G)$.

Q.E.D.

Every solvable group has a LIM on $B(G)$ since it is obtained by successive extensions of amenable groups by abelian groups (use 1.2.6). No group which contains a free group on two generators can be amenable. For many other examples and some extensions of these results to semigroups, Day [8] and Dixmier [11] are quite comprehensive; [11] discusses groups whose generators $\{x_i : i \in J\}$ have only relations of the form $(x_i^{r_i} = e)$. He shows that G fails to be amenable unless the index set J has only one element (G singly generated abelian), or J has two elements and the relations have the form $x_1^2 = x_2^2 = e$. Although it is not mentioned explicitly, one can demonstrate the presence of a free group on two generators within all the other groups; Dixmier proved non-amenability by appealling to criterion (D). This observation is compatible with the following conjecture:

Conjecture. A discrete group G fails to be amenable $\Longleftrightarrow$ there exists a free group on two generators within G.

This is decidedly an unsolved problem (($\Longleftarrow$) is trivial). When we consider means on topological groups we shall exhibit some remarkable results on connected groups which support this conjecture.

Example 1.2.8. There is a free group on two generators in the proper orthogonal group $G = SO(3, \mathbf{R})$, so G fails to be amenable (when regarded as a discrete group).

Proof. In the matrix group $SL(2, \mathbf{C})$ the subgroup $SU(2) = \{A: AA^* = I = A^*A, \det(A) = 1\}$ consists of all matrices:

$$\begin{pmatrix} a & b \\ c & d \end{pmatrix} \text{ with } d = \bar{a},\ c = -\bar{b},\ |a|^2 + |b|^2 = 1,$$

and gives a two-fold covering group for $SO(3) = SO(3, \mathrm{R})$ when the rotations of the unit sphere in R^3 are realized as fractional linear transformations of C via stereographic projection and we identify a 2×2 matrix with the transformation $z \to \dfrac{az+b}{cz+d}$.

The groups $G = SO(3),\ SU(2),\ SL(2, \mathrm{C})$ admit no relations of the form

$$(*) \qquad \sigma^{i_1} \tau^{j_1} \cdots \sigma^{i_m} \tau^{i_m} = e, \quad \text{all } (\sigma, \tau) \in G \times G$$

where $(i_1, \ldots, i_m)$ and $(j_1, \ldots, j_m)$ are integers with $i_k \neq 0$, $j_k \neq 0$ except for i_1 and j_m if $m \geq 2$ (except for i_1 or j_m if $m = 1$). To see this we first show that there can be no such relation in $SL(2, \mathrm{C})$. In fact, consider α: $z \to z + 2$ and ω: $z \to -1/z$ in $SL(2, \mathrm{C})$; if we can show that there is no relation of the form $\omega \alpha^{p_1} \omega \cdots \omega \alpha^{p_m} = e$ ($p_k \neq 0$ except possibly p_m, $m \geq 1$) between these elements, then the elements $\sigma = \omega\alpha\omega$, $\tau = \alpha$ cannot satisfy any relation like (*). Assuming such a relation to hold we would have:

$$z = \cfrac{-1}{2p_1 - \cfrac{\ddots}{\cfrac{1}{2p_{m-1} - \cfrac{1}{2p_m + z}}}}, \quad \text{all } z \in \mathrm{C}.$$

If we let $|z| \to \infty$ the left side approaches $+\infty$ in modulus while the right side stays bounded by 1 since we have:

$$|a_{m-1}| = |2p_{m-1}| \geq 2, \quad |a_{m-2}| = \left|2p_{m-2} - \frac{1}{a_{m-1}}\right| \geq 3/2, \ldots,$$

$$|a_1| = \left|2p_1 - \frac{1}{a_2}\right| \geq \frac{m}{m-1} \geq 1.$$

[Since a, ω actually correspond to matrices in $SL(2, \mathrm{R})$ we have also shown that $SL(2, \mathrm{R})$ has no relations of form (*)]. Now $SU(2)$ is a closed analytic submanifold in $SL(2, \mathrm{C})$ whose Lie algebra $su(2) = \{A\colon A^* + A = 0,\ A^{tr} + A = 0\}$ (A^{tr} = transpose, A^* = conjugate of A^{tr}), is a real form for the Lie algebra $sl(2, \mathrm{C}) = \{A\colon A^{tr} + A = 0\}$; i.e., the complex vector space $sl(2, \mathrm{C})$ decomposes as a direct sum of real vector spaces: $sl(2, \mathrm{C}) = su(2) \oplus i\ su(2)$. Suppose there is a relation (*) with $F(\sigma, \tau) = \sigma^{i_1} \tau^{j_1} \cdots \sigma^{i_m} \tau^{j_m} = e$ for all

$$(\sigma, \tau) \in SU(2) \times SU(2) \subset SL(2, \mathrm{C}) \times SL(2, \mathrm{C});$$

we assert that we must then have $F(\sigma, \tau) = e$ for all $\sigma, \tau \in SL(2, \mathrm{C})$, which is a contradiction, so $SU(2)$ admits no non-trivial relations. To prove this, define

$$G(S, T) = \log(F(\mathrm{Exp}(S), \mathrm{Exp}(T)))$$

for all $S, T \in sl(2, \mathrm{C})$ near $(0, 0)$; then $G(S, T)$ is real analytic near $(0, 0) \in sl(2, \mathrm{C}) \times sl(2, \mathrm{C})$ and $G(S, T) \equiv 0$ on the real subspace $su(2) \times su(2)$. Let $N = \dim_R(su(2))$; if we regard $sl(2, \mathrm{C})$ as an N-dimensional vector space over C, G becomes a complex analytic function in $2N$ complex variables $z_1, \ldots, z_{2N}$ which vanishes near $(0, \ldots, 0)$ on all points with $\mathrm{Im}(z_1) = \cdots = \mathrm{Im}(z_{2N}) = 0$. By Cauchy-Riemann equations we see that $G \equiv 0$ near $(0, \ldots, 0)$, hence $F(\sigma, \tau) \equiv e$ for all (σ, τ) near the unit in $SL(2, \mathrm{C}) \times SL(2, \mathrm{C})$, which $\Longrightarrow F(\sigma, \tau) \equiv e$ throughout the connected analytic group $SL(2, \mathrm{C}) \times SL(2, \mathrm{C})$. Since $SU(2)$ is a twofold covering group for $SO(3)$ it is clear that $SO(3)$ admits no relation of the form (*).

Let $N = 9$; then the matrix group $SO(3)$ appears as a real analytic submanifold in R^N and if we write $\sigma = (\sigma_1, \ldots, \sigma_N)$

for $\sigma \in SO(3)$ then $((\sigma\tau)_1, \ldots, (\sigma\tau)_N)$ has coefficients which are polynomials with rational coefficients in the $2N$ variables $\sigma_1, \ldots, \tau_N$. If $P(x_1 \cdots x_{2N})$ is any polynomial in $2N$ variables with rational coefficients, then $P \equiv 0$ for all $(x, y) \in SO(3) \times SO(3)$ near some point $(x', y') \in SO(3) \times SO(3) <=> P \equiv 0$ throughout $SO(3) \times SO(3)$ since P gives an analytic function on this connected submanifold. Let $J = \{P\colon P \equiv 0 \text{ on } SO(3) \times SO(3)\}$, an ideal is the algebra A of all polynomials in $2N$ variables with rational coefficients. There must exist points $\sigma, \tau \in SO(3)$ such that $P(\sigma, \tau) \neq 0$ for all $P \in A \backslash J$ (of course $P(\sigma, \tau) = 0$ if $P \in J$), for there are only denumerably many $P \in A$ and $P \in A \backslash J => P \not\equiv 0$ on $SO(3) \times SO(3)$, and since P is analytic on the manifold $SO(3) \times SO(3)$ its zero set N_P must be nowhere dense in the complete metric space $SO(3) \times SO(3)$. Thus we cannot have $SO(3) \times SO(3) = \cup \{N_P\colon P \in A \backslash J\}$. Pick some (σ, τ) with this algebraic independence property. If there is some relation $\sigma^{i_1} \tau^{j_1} \cdots \sigma^{i_m} \tau^{j_m} = e$ then in terms of the coefficients σ_i, τ_i and the polynomials $\{P_k \in A\colon k = 1,2 \ldots N\}$ describing group multiplication, $P_k(\sigma_1, \ldots, \sigma_N, \tau_1, \ldots, \tau_N) = 0$. By our choice of σ, τ, this $=> P_k \in J$ so $P_k(\sigma', \tau') = 0$ for all $(\sigma', \tau') \in SO(3) \times SO(3)$, and we have the non-trivial relation $\sigma^{i_1} \tau^{j_1} \cdots \sigma^{i_m} \tau^{i_m} = e$ for all $(\sigma, \tau) \in SO(3) \times SO(3)$, which is a contradiction. Thus the elements (σ, τ) we have chosen generate a free group in $SO(3)$. Q.E.D.

Remark. I do not know whether non-trivial relations of form (*) can exist in a connected semi-simple Lie group. If not, then all such groups will evidently include free groups, since Lie groups are locally isomorphic to matrix groups and the latter part of our discussion applies. Part of this argument appeared in [72].

§1.3. VON NEUMANN'S WORK ON INVARIANT MEASURES

In [72] von Neumann considered the following problem: if G is a group acting as a transformation group on a set S, so $g_1 g_2(x) = g_1(g_2(x))$ and $e(x) = x$ (e the unit in G), we are to find a normalized "measure" on S which is G-invariant. Even if $S = [0, 1]$ and G the circle group acting by translation (mod 1) there cannot be a non-trivial σ-additive measure on $\Omega(S)$–all subsets–which is normalized so $\mu(S) = 1$. We ask therefore only for a G-invariant finitely additive measure $\mu\colon \Omega(S) \to [0, +\infty]$, so $\mu(\emptyset) = 0$ and:

(i) $\mu(\cup_{i=1}^N E_i) = \sum_{i=1}^{N} \mu(E_i)$ for disjoint $\{E_i\}$ in Ω;

(ii) $\mu(gE) = \mu(E)$ all $E \in \Omega$, $g \in G$,

which is normalized on some set $A \subset S$, $A \neq \emptyset$;

(iii) $\mu(A) = 1$.

We refer to such a measure as an invariant measure for the system (G, S, A). Several examples attracted attention, particularly:

Example 1. $S = \mathbf{R}^n$, $A = \{x\colon 0 \leq x_i \leq 1$ for $i = 1, 2, \ldots, n\}$, G all isometries of $\mathbf{R}^n$.

Example 2. $S = K^n = \{x \in \mathbf{R}^n\colon \Sigma_{i=1}^n x_i^2 = 1\}$, $A = S$, and G the isometries of $\mathbf{R}^n$ leaving K^n fixed.

Hausdorff [32] showed that an invariant measure fails to exist in Example 2 for $n = 3$, from which one can easily see that an invariant measure fails to exist in Example 1 for $n = 3$, however, Banach [1] showed that such measures exist for $n = 1, 2$ in Example 1. It was von Neumann's insight that the difference between dimensions $n = 1, 2$ and $n \geq 3$ can be traced to profound differences in the algebraic properties of groups G rather than differences in the transformation spaces $S = \mathbf{R}^n$, K^n; specifically, a free group on two generators appears as a subgroup of G for $n \geq 3$ but not for $n = 1, 2$.

Note: The early investigations in [32] were generalized by Banach and Tarski in [2] and the family of phenomenae they exhibit are referred to as the *Hausdorff-Banach-Tarski paradox*. Hausdorff's example shows in an extremely dramatic way that an invariant measure cannot exist in Example 2 for $n = 3$. The relevant example, whose proof we shall not attempt, shows that K^3 can be partitioned into four disjoint sets A, B, C, D such that

(1) D is invariant, denumerable, and may be shown to have zero measure relative to any finitely normalized invariant measure on K^3.

(2) There is some $120°$ rotation σ whose iterates carry the pieces A, B, C onto one another.

(3) There is some $180°$ rotation τ which carries each of the pieces A, B, C onto the union of the other two.

If an invariant measure existed, each of the pieces A, B, C (D is irrelevant) would have to carry, simultaneously, both $\frac{1}{2}$ and $\frac{1}{3}$ of the total mass (compare with Exercise 1.2.3 which exhibits a primitive version of this paradox in the free group on two generators).

The problem of finding an invariant measure for the system (G, G, G) is equivalent to finding a LIM for $B(G)$; if LIM m is given, take $\mu(E) = \langle m, \chi_E \rangle$ where χ_E is the characteristic function of E, and conversely if μ is an invariant measure (finitely additive) for our system we define

$$\overline{m}\left(\sum_{i=1}^{N} a_i \chi_{E_i}\right) = \sum_{i=1}^{N} a_i \mu(E_i)$$

if E_i are disjoint. This is a non-negative G-invariant linear functional on the submanifold B_0 of all simple functions in $B(G)$ and is continuous in the $\|\cdot\|_\infty$-norm because it is non-negative and $1 \in B_0$. Since B_0 is norm dense in $B(G)$, $\overline{m}$ extends by continuity to the desired LIM on $B(G)$. More generally consider a system (G, S, A), define the *bounded sets* in S to be those covered by a finite union of translates $\{gA: g \in G\}$, and let $X = \{f \in B(S): \text{supp}(f) \text{ is bounded}\}$.

Proposition 1.3.1. Let G be amenable. Then there is an invariant measure for $(G, S, A) \Longleftrightarrow$ there is a non-negative G-invariant linear functional m on X with $< m, \chi_A > = 1$.

Note. S may not be bounded, so we may have $\mu(E) = +\infty$ for certain sets in S. If $A = S$, m is just the G-invariant *mean* on $X = B(S)$ constructed above.

Proof. Given m, if $E \subset S$ is bounded let $\mu(E) = < m, \chi_E >$ and otherwise $\mu(E) = +\infty$, proving $(\Longleftarrow)$. Conversely let X_0 be the submanifold of *simple* functions with bounded support and define $\overline{m}(\Sigma_{i=1}^N a_i \chi_{E_i}) = \Sigma_{i=1}^N a_i \mu(E_i)$ if E_i are disjoint bounded sets. If $E \subset S$ is bounded, then $X_E = \{f \in X: \text{supp}(f) \subset E\}$ is complete in the sup-norm, contains X_E^0 = the simple functions with support in E as a norm dense submanifold, and $\overline{m} \mid X_E^0$ extends by continuity to a non-negative linear functional m_E on X_E. But if $E \subset F$ it is evident that m_F extends m_E so we may define $m(f) = m_E(f)$ for any bounded set $E \supset \text{supp}(f)$ to get the desired G-invariant functional m on X. Q.E.D.

Within X define the submanifold Y spanned by $\{\chi_{gA}: g \in G\}$. We can use a variant of the Hahn-Banach theorem to get a simple criterion for existence of an invariant measure in (G, S, A) when G is known to be amenable.

Theorem 1.3.2. Let G be an amenable group. Then there is an invariant measure for the system (G, S, A) $\Longleftrightarrow$ there is a non-negative G-invariant linear functional ϕ on Y with $< \phi, \chi_A > = 1$.

Proof. Implication ($\Longrightarrow$) is immediate from 1.3.1; the converse follows if we can extend ϕ to a non-negative G-invariant linear functional on X. We construct this extension following Dixmier [11].

First, suppose ϕ may be extended to a non-negative linear functional ψ on X, *not* necessarily G-invariant; we construct a G-invariant functional Ψ as follows. Let $f \in X$ and define $\hat{f}(x) = < \psi, L_x f >$ where $L_x f(y) = f(xy)$; $\hat{f}$ is bounded for if $\text{supp}(f) \subset \cup_{i=1}^N g_i A$, then $|f| \leq \|f\|_\infty \cdot \Sigma_{i=1}^N \chi_{g_i A} = h \in Y$ and so $|\psi({}_x f)| \leq |\psi({}_x h)| = |\phi({}_x h)| = |\phi(h)|$ by G-invariance of ϕ on Y. Applying any LIM m on $B(G)$, we define $\Psi(f) = < m, f\hat{\ } > = < m(x), \psi(L_x f) >$, which is a G-invariant non-negative functional on X since $({}_t f)\hat{\ } = {}_t(f\hat{\ })$. Thus modulo the extension of ϕ to ψ on X, we have proved ($\Longleftarrow$).

Let $K = \{f \in X: f \geq 0\}$, a convex cone; if $f \in K \cap Y$ then $f \geq 0 \Longrightarrow \phi(f) \geq 0$ so $K \cap Y$ lies to one side of the hyperplane $N = \mathrm{Ker}(\phi)$ in Y. Moreover, if $f \in X$ there is some $g \in Y \cap K$ such that $\lambda f + g \in K$ for all sufficiently small $\lambda \in \mathbf{R}$; in fact if $\mathrm{supp}(f) \subset \cup_{i=1}^{N} g_i A$ we may take

$$g = 2\|f\|_\infty \sum_{i=1}^{N} \chi_{g_i A}$$

and are assured $\lambda f + g \geq 0$, so $\lambda f + g \in K$, for $|\lambda| < 1$. Existence of ψ on X extending ϕ is immediate from the following variant of the Hahn-Banach theorem. Q.E.D.

Lemma 1.3.3. (Dixmier [11].) Let $Y \subset X$ be real vector spaces. Let $K \subset X$ be a convex set with the property: if $f \in X$ there is some $g \in K \cap Y$ such that $\lambda f + g \in K$ for all small λ. Each linear functional f on Y which is non-negative on $K \cap Y$ extends to a linear functional on X which is non-negative on K.

The proof can be found in Dixmier [11], p. 226.

Examples. If $S = A$ then Y is the space of all constant functions, $X = B(S)$, and the hypotheses of 1.3.2 are trivially satisfied, thus there is an invariant measure for any system (G, S, S) if G is amenable, and such measures exist for the systems in Example 2 if $n = 1, 2$ since the group G_n is abelian. We have seen that the 3-dimensional rotation group $SO(3, \mathbf{R})$ includes a free group on two generators, so the G_n fail to be amenable for $n \geq 3$; evidently 1.3.2 *does not* resolve this existence question for $n \geq 3$.

Consider the systems (G_n, S, A) of Example 1. Then Y consists of all functions $f = \sum_{i=1}^{N} \lambda_i \chi_{x_i A}$ and each of these is Borel-measurable; thus we get a G_n-invariant non-negative linear functional on Y from Lebesgue measure dt in $S = \mathbf{R}^n$:

$$< \phi, f > = \int_{\mathbf{R}^n} f(t)dt \quad .$$

If dt is normalized as usual: $< \phi, \chi_A > = 1$, so 1.3.2 applies. For $n = 1, 2$; (but not $n \geq 3$) G_n is amenable (solvable for $n = 1, 2$; includes $SO(3, \mathrm{R})$ for $n \geq 3$); again, 1.3.2. fails to resolve the problem if $n \geq 3$.

We have seen that amenability of G, and simple restrictions on the action of G, imply the existence of an invariant measure for (G, S, A). It is not clear to what extent amenability of G is *necessary* for existence of such measures; specifically, if G is non-amenable, and acts reasonably on S, can there be an invariant measure for (G, S, A)? Even in the simple case $S = A$ (where we know amenability of G implies existence of such a measure) this converse question is not well resolved.

Problem. If there is an invariant measure for a system (G, S, S), and if G acts reasonably on S, is G amenable?

Some restriction on the action of G must be made. For example if G acts trivially: $x(s) = s$ for all $x \in G$, $s \in S$, we could take any mean on $B(S)$ and get an invariant measure; presumably we want to require that the action have "trivial kernel" so the unit is the only element $x \in G$ with $x(s) = s$ all $s \in S$. The principal result of [72] in this direction shows that there is no invariant measure for (G, S, A) if G is non-amenable in a particularly horrible way.

Theorem 1.3.4. Invariant measures fail to exist for (G, S, A) if there is a subset $A' \subset A$ and a sequence of free groups in G

$\{F_N : N = 1, 2\ldots\}$ on two generators, a_N and b_N respectively, with properties:

(1) There are translations $\{g_1 \cdots g_n\}$ such that

$$A \subset \bigcup_{i=1}^{n} g_i A' .$$

(2) For each $N = 1, 2\ldots$ we have

$$a_N^{i_1} b_N^{j_1} \cdots a_N^{i_m} b_N^{j_m} A' \subset A$$

for all m-tuples of integers $(i_1, \ldots, i_m)$ and $(j_1, \ldots, j_m)$ with $|i_1| + \cdots + |i_m| + |j_1| + \cdots + |j_m| \leq N$.

If $x = e$ is the only element such that $x: S \to S$ has a fixed point this is all we need; however if G does not act freely in this sense we must also require:

(3) For some fixed integer M, the joint action of $F_1, \ldots, F_M$ on S is free: i.e., if $s \in S$ and $g_1 \in F_1, \ldots, g_M \in F_M$ are given with $g_i \neq e$ for $i = 1, 2 \ldots M$, then we cannot have $g_i(s) = s$ for all $i = 1, 2 \ldots M$.

One can extend the discussion of 1.2.8 to show that these conditions are satisfied for $n \geq 3$ in Example 1, Example 2 (condition (3) must be verified since these groups obviously do not act freely; (1) and (2) are trivial if $S = A$, as in Example 2). We shall not reproduce the proof; the general idea is that conditions (1)...(3) insure that an analog of the Hausdorff-Banach-Tarski paradox may be constructed.

Here are a few suggestive examples concerning the problem set forth above.

Example 1.3.5. Let G act transitively and *freely* on S: each $x \neq e$ in G is fixed point free. Then there is an invariant measure for (G, S, S) <=> G is amenable. We have already seen (<=); conversely, fix $s_0 \in S$ and lift $f \in B(G)$ to $f^\wedge \in B(S)$ by defining $f^\wedge(g(s_0)) = f(g)$, all $g \in G$. There is a G-invariant mean m on $B(S)$ so we get a LIM for $B(G)$ by taking $< \overline{m}, f > = < m, f^\wedge >$, since $({}_g f)^\wedge(s) = f^\wedge(g^{-1} s)$.

Example 1.3.6. Consider the system (G, S, S) in Example 2 fo $n \geq 3$ (we do $n = 3$; proof generalizes with some technical co plications). If there is an invariant measure for this system may induce a LIM on $B(G)$ by the following construction. Le $p \in S$ be fixed, say $p = (1, 0, 0)$, and $H_p = \{g \in SO(3)\colon g(p) =$ the (abelian) group of rotations about the x-axis. Let $SO(3)/$ be the space of left cosets $\{xH_p : x \in SO(3)\}$, let m_1 be a LI on $B(H_p)$, and define $\psi f(x) = < m_1, ({}_x f | H_p) >$ for $f \in B(SO(3)$ then $\psi\colon f \to \psi f$ is a linear map from $B(SO(3))$ to the submani fold B^* of functions constant on left cosets of H_p with $\psi(1)$ 1, $f \geq 0 \Rightarrow \psi f \geq 0$, and $\psi({}_x f) = \psi({}_{hx} f) = {}_x(\psi f)$ for $x \in SO(3)$, $h \in H_p$. Now assume there is a G-invariant mean m_2 for $B(S)$. For $x, y \in SO(3)$, we have $x(p) = y(p)$ in S <=> $xH_p = yH_p$ and since $SO(3)$ is transitive on S we get a natur surjection $\lambda\colon SO(3) \to S$ by letting $\lambda(xH_p) = x(p)$. Then w may lift $B^* \subset B(G)$ linearly into $B(S)$ by defining $\lambda^*\colon B^* \to B(S)$ so $\lambda^* f(x(p)) = f(xH_p) = f(x)$; evidently $\lambda^*(1) = 1$, $f \geq 0 \Rightarrow \lambda^* f \geq 0$, and $\lambda^*({}_y f) = {}_y(\lambda^* f)$. For $f \in B(SO(3))$ w set $< m, f > = < m_2, \lambda^* \circ \psi(f) >$ and get a LIM on $SO(3)$.

We have seen that $SO(3)$ is *not* amenable, so there cannot be an invariant mean for (G, S, S) in Example 2.

SECTION 2

INVARIANT MEANS ON LOCALLY COMPACT GROUPS

§2.1 VARIOUS DEFINITIONS OF INVARIANT MEANS ON TOPOLOGICAL GROUPS

If one considers means on $X = B(G)$ this amounts to studying G as a discrete group. Means on this function space will be of peripheral interest for non-discrete locally compact groups. The spaces X which have attracted current interest reflect the topological structure of G; we shall discuss left invariant means on

$X = L^\infty(G)$, essentially bounded Borel measurable functions
$X = CB(G)$, bounded continuous functions
$X = UCB_r(G)$, right uniformly continuous bounded functions
$X = UCB(G)$, two-sided uniformly continuous bounded functions.

Here f is *right uniformly continuous* if, given $\varepsilon > 0$, there is a neighborhood $U(\varepsilon)$ of the unit in G such that:

$$|f(x) - f(yx)| < \varepsilon \quad \text{all } x \in G, \quad y \in U(\varepsilon).$$

Left uniform continuity is similarly defined and we have $UCB = (UCB_r) \cap (UCB_\ell)$. Each of the above spaces is two-sided translation invariant and is a sup-norm closed subspace in the preceding one.

Remark: In this chapter we deal exclusively with groups, and use the notation ${}_xf(t) = f(x^{-1}s)$. All groups will be locally compact. Some of our remarks can be applied to topological groups and semigroups.

Remark: This notion of mean on $L^\infty(G)$, and even the definition of $L^\infty(G)$, require some comment for non-separable groups (see [34], section 11 for details). We define $L^\infty(G)$ to be the bounded Borel measurable functions (Borel sets the σ-algebra generated by closed sets, as in [34]) and, with respect to left Haar measure in G, define the locally null sets. We identify functions which differ only on a locally null set and make $L^\infty(G)$ a Banach algebra with norm

$$\|f\|_\infty = \text{ess sup}\{|f(t)| : t \in G\}$$

$$= \inf\{\sup\{|f(t)|: t \in G\backslash N\}: N \text{ a locally null set in } G\}\,.$$

Means in $L^\infty(G)$ are defined as in section 1.1, substituting ess sup (ess inf) for sup (inf) in condition (2). In this context we consider $L^1(G)$, the Borel measurable functions which are integrable with respect to left Haar measure, taken with the usual norm $\|f\|_1 = \int |f(t)|dt$. Then $L^\infty(G) = (L^1)^*$ and the functions $P(G) = \{f \in L^1(G): f \geq 0, \int |f(t)|dt = 1\}$ form a weak-$*$ dense convex subset in the weak$*$-compact convex set Σ of all means on $L^\infty(G)$, by arguments similar to those which give density of the discrete means.

Let $C_0(G)$ be the continuous functions vanishing at infinity and $M(G) = (C_0(G))^*$, the Banach space of bounded regular Borel measures on G with total variation norm. We shall write $<\mu, f> = \int f d\mu$ for $\mu \in M(G)$, $f \in C_0(G)$. In $M(G)$ we have the natural involution (isometric, anti-isomorphic) $\mu \to \mu^*$ defined by

$$< \mu^*, \psi > = \overline{\left(\int \overline{\psi(s^{-1})}\, d\mu(s) \right)} \quad \text{all } \psi \in C_0(G),$$

and $M(G)$ becomes a Banach algebra under the convolution operation

$$< \mu * \nu, \psi > = \int\int \psi(st) d\mu \times \nu(x, t) \quad \text{all } \psi \in C_0(G).$$

Let dt be a fixed *left* Haar measure on G, then $L^1(G, dt)$ is isometric and isomorphic with the two sided ideal in $M(G)$ of all measures absolutely continuous relative to the Haar measure if we identify f with μ_f defined so $d\mu_f(t) = f(t)dt$. It is easily verified that the natural involution appears in $L^1(G) \subset M(G)$ as

$$d[\mu_f{}^*](t) = \overline{f(t^{-1})}\, \Delta(t^{-1}) dt$$

where Δ is the modular function on G. For $\mu \in M(G)$, $f \in L^p(G)$ with $1 \leq p < \infty$, we define

$$\mu * f(s) = \int f(t^{-1} s)\, d\mu(t)$$

$$f * \mu(s) = \int f(st^{-1})\, \Delta(t^{-1})\, d\mu(t)$$

which $\Longrightarrow$ $\|\mu * f\|_p \leq \|\mu\| \cdot \|f\|_p$ and $\|f * \mu\|_p \leq \|\mu\| \cdot \|f\|_p$. When $\phi \in L^1(G)$ is identified with a measure as above, this gives the usual formula

$$\phi * f(s) = \mu_\phi * f(s) = \int f(t^{-1} s)\, \phi(t) dt$$

$$f * \phi(s) = f * \mu_\phi(s) = \int f(t)\, \phi(t^{-1} s) dt \quad .$$

For $x \in G$ write δ_x for the point mass at x; then if $f \in L^p(G)$:

$$\delta_x * f(s) = f(x^{-1}s) = {}_x f(s)$$

$$f * \delta_x(s) = f(sx^{-1})\Delta(x^{-1}).$$

If $f \in L^\infty(G)$ there are special difficulties when G is not unimodular ($\phi \in L^1$ does not $\Rightarrow \phi^\sim \in L^1$); the convolutions which make sense are:

$$\phi * f(s) = \int f(t^{-1}s)\phi(t)dt$$

$$f * \phi^\sim(s) = \int f(t)\phi^\sim(t^{-1}s)dt = \int f(t)\phi(s^{-1}t)dt$$

where $\phi \in L^1(G)$. It is easy to verify $\|\phi * f\|_\infty \leq \|\phi\|_1 \cdot \|f\|_\infty$ and $\|f * \phi^\sim\|_\infty \leq \|\phi\|_1 \cdot \|f\|_\infty$. If G is not unimodular, $f * \phi$ need not make sense.

Hulanicki made very effective use of the notion of a topologically left invariant mean on L^∞ in [37]. His definition also makes sense for $X = CB(G)$, $UCB_r(G)$, and $UCB(G)$.

Definition 2.1.1. A linear functional m on X is a *topologically left invariant mean* if m is a mean and $m(\phi * f) = m(f)$ for $f \in X$, whenever $\phi \in P(G) = \{\phi \in L^1(G)\colon \phi \geq 0, \|\phi\|_1 = 1\}$. It is *topologically right invariant* if $m(f * \phi^\sim) = m(f)$ for all $\phi \in P(G)$, $f \in X$.

The following easily verified lemma justifies this definition for $X = UCB_r(G)$ and $UCB(G)$, and is essential to later discussions.

Lemma 2.1.2. If $f \in L^\infty(G)$ and $\phi \in P(G)$ then $\phi * f \in UCB_r(G)$ and $f * \phi^\sim \in UCB_\ell(G)$. If $g \in UCB_\ell(G)$ then $\phi * g \in UCB(G)$, and if $g \in UCB_r(G)$ then $g * \phi^{\sim} \in UCB(G)$.

Hulanicki [37] has also shown (the proof works equally well for the other function spaces in our list, and for right invariance):

Proposition 2.1.3. If m is a topological LIM on $L^{\infty}(G)$ then m is also a LIM on $L^{\infty}(G)$.

Proof: Fix $\phi \in P(G)$. Then it is trivial that

$$\phi * ({}_x f) = \Delta(x^{-1})(\phi_{x^{-1}}) * f,$$

thus:

$$m({}_x f) = m(\phi * ({}_x f)) = m((\Delta(x^{-1})\phi_{x^{-1}}) * f) = m(f)$$

since $\Delta(x^{-1})\phi_{x^{-1}} \in P(G)$ again. Q.E.D.

Simple modifications of the considerations in 1.2 show that the obvious analogs of Dixmier's criterion (D) for existence of a LIM in $X = B(G)$ are valid for each of the function spaces X in our list and for topological left (right) invariance as well as left (right) invariance; for example there is a LIM on $L^{\infty}(G)$ if and only if

$$\text{ess inf}\left\{\sum_{i=1}^{N} (f_i - {}_{x_i}f_i)\right\} \leq 0,$$

all real valued $\{f_1, \ldots, f_N\}$ in $L^{\infty}(G)$ and $\{x_1, \ldots, x_N\} \subset G$, $N < \infty$. Likewise there is a topological LIM on $L^{\infty}(G)$ if and only if

$$\text{ess inf}\left\{\sum_{i=1}^{N} (f_i - \phi_i * f_i)\right\} \leq 0$$

for $\{f_1 \ldots f_N\}$ real valued in L^{∞}, $\{\phi_1 \ldots \phi_N\} \subset P(G)$, and $N < \infty$.

§2.2. EQUIVALENCE OF VARIOUS TYPES OF INVARIANT MEANS

The study of topological left invariant means on $L^\infty(G)$ in [37] was an important step in clarifying the relation between existence of a left invariant mean and the weak containment property mentioned in the introduction. To relate these efforts to other results which deal with ordinary invariant means it is important to know that the converse of 2.1.3 is true: to be precise we need to know whether existence of a LIM on $L^\infty(G)$ implies existence of a topological LIM on $L^\infty(G)$. The main theorem, valid for any locally compact group, resolves this and many related questions.

Theorem 2.2.1. Existence of a LIM on $B(G)$ implies existence of a LIM on $CB(G)$ but the converse may fail to be true. Each of the following properties implies the others:

(1) there is a topological LIM on $L^\infty(G)$;

(2) there is a LIM on $L^\infty(G)$;

(3) there is a LIM on $CB(G)$;

(4) there is a LIM on $UCB_r(G)$;

(5) there is a LIM on $UCB(G)$.

Proof: As $CB(G)$ is a closed subspace of $B(G)$, mere restriction of a LIM m on $B(G)$ to $CB(G)$ gives a LIM $\tilde{m}$ on $CB(G)$. Notice $\tilde{m}$ is non-trivial since $\tilde{m}(1) = m(1) = 1$. Conversely, it is clear that the (compact) real 3-dimensional orthogonal group $O(3)$ in its usual Lie group topology has a LIM on $CB(G)$, but we have seen that $B(G)$ fails to have a LIM because G has a (non-closed) subgroup isomorphic to the free group on two generators.

We get (1) => (2) from 2.1.3 and by restriction we see (2) => ⋯ => (5) is a triviality. Namioka [55] has given an elementary proof that (2) => (1); his argument can be used to prove much more(2):

Lemma 2.2.2. If m is a LIM on $UCB(G)$, then m is also a topological LIM on $UCB(G)$.

Proof: Let $f \in UCB(G)$ and $\phi \in L^1(G)$. Left invariance of m gives

$$m(({}_x\phi) * f) = m({}_x(\phi * f)) = m(\phi * f)$$

and $\phi * f \in UCB(G)$ by 2.1.2, thus $\phi \to m(\phi * f)$ is a left invariant bounded functional on $L^1(G)$ and so there exists a constant $k(f)$ such that

$$(*) \qquad m(\phi * f) = k(f) \int \phi(t)dt \quad \text{all } \phi \in L^1(G).$$

If $\phi \in P(G)$, so $\int \phi(t)dt = 1$, we have $m(\phi * f) = k(f)$. Now let $\{e_j\} \subset P(G)$ be an approximate identity for $L^1(G)$; it is readily verified that $f \in UCB_r(G) \Rightarrow \|e_j * f - f\|_\infty \to 0$, so that (*) implies:

(2) The author proved 2.2.2 and used it to prove (5) =>(1) by techniques quite different from those in the (then unpublished) paper [55], and belatedly learned of Namioka's elegant proof of (2) =>(1) in [55]. In subsequent correspondence Namioka pointed out the argument presented here which is much simpler than the author's original proof. E. Granirer, in an independent effort, has recently proved an analog of 2.2.2 which is valid for general topological groups—see [25]. The first proof of 2.2.1 (in slightly weaker form) appeared in Reiter [64]; Reiter's proof was much less direct than the present one.

$$m(\phi * f) \leftarrow m((\phi * e_j) * f) = k(f) = m(e_j * f) \to m(f)$$

for all $\phi \in P(G)$ (note $\phi * e_j \in P(G)$ too). Q.E.D.

The proof applies unaltered to $UCB_r(G)$. It is not known whether 2.2.2 is true for $CB(G)$ or $L^\infty(G)$. We give another proof of this key lemma, based on the vector-valued integral formula $\phi * f = \int \phi(x)_x f\, dx$, in Appendix 3.

To see (5) $\Longrightarrow$ (1) let m be a LIM on $UCB(G)$, $E = E^{-1}$ a compact neighborhood of the unit in G, and consider the normalized characteristic function $\phi_E = \tilde{\phi}_E \in P(G)$. Evidently (see 2.1.2) $\phi_E * f * \phi_E \in UCB(G)$ for all $f \in L^\infty(G)$ and $\tilde{m}(f) = m(\phi_E * f * \phi_E)$ is a mean on $L^\infty(G)$. It is actually a topologically left invariant mean; indeed, consider any $g \in UCB_\ell(G)$ (such as $g = f * \phi_E$ above), let $\lambda_1, \lambda_2 \in P(G)$, and let $\{e_n\} \subset P(G)$ be an approximate identity for $L^1(G)$. Then $\|\lambda_i * e_n - \lambda_i\|_1 \to 0$, which $\Longrightarrow$ $\|\lambda_i * e_n * g - \lambda_i * g\|_\infty \to 0$, and by 2.1.2 we have $e_n * g \in UCB(G)$ for all e_n. Now topological left invariance of m on $UCB(G)$ gives:

$$\begin{aligned} m(\lambda_1 * g) \leftarrow m(\lambda_1 * (e_n * g)) &= m(e_n * g) \\ &= m(\lambda_2 * (e_n * g)) \to m(\lambda_2 * g) \end{aligned}$$

so that $m(\lambda_1 * g) = m(\lambda_2 * g)$, all $g \in UCB_\ell(G)$ and all $\lambda_i \in P(G)$. But $f \in L^\infty \Longrightarrow f * \phi_E \in UCB_\ell(G)$: hence $\tilde{m}$ is topologically left invariant:

$$\tilde{m}(\phi * f) = m(\phi_E * \phi * (f * \phi_E)) = m(\phi_E * f * \phi_E) = \tilde{m}(f)$$

for all $\phi \in P(G)$, proving the main theorem. Q.E.D.

In view of 2.2.1 it is natural to say that a locally compact group is *amenable* if any one of the properties (1)...(5) holds. This should really be called left amenability, but our terminology will cause no confusion.

Let G be an amenable locally compact group. The existence of two-sided invariant means on the spaces in our list is not so clear as it is for discrete groups (see 1.1.3). Using the equivalence theorem we show that there are two-sided (topologically) invariant means on each space; it is evidently sufficient to produce a topologically invariant mean m on L^∞, so $m(\phi * f) = m(f * \phi^\sim) = m(f)$ for all $\phi \in P(G)$. Exactly as in (2.1.3) we see that topological invariance implies invariance; notice that for each of our spaces: $f \in X \Longrightarrow \phi * f$ and $f * \phi^\sim$ are in X.

First we produce a two-sided invariant mean m for UCB: let m_ℓ be a LIM on $UCB(G)$, m_r a right invariant mean on $CB(G)$, and define $m(f) = \langle m_r, F \rangle$ where $F(x) = \langle m_\ell, f_x \rangle$. Here $x \to x_0$ in $G \Longrightarrow \|f_x - f_{x_0}\|_\infty \to 0$ for $f \in UCB$, so $F \in CB(G)$; it is readily verified that m is two-sided invariant, hence topologically left invariant by (2.2.2). Trivial modifications of (2.2.2) show that m is topologically right invariant. Next define $\tilde{m}(f) = m(\phi_E * f * \phi_E)$ for $f \in L^\infty(G)$, as in the proof of (5) $\Longrightarrow$ (1); from those considerations we see that $\tilde{m}$ is a topological LIM and trivial modifications show topological right invariance.

§2.3. COMBINATORY PROPERTIES OF AMENABLE GROUPS

When we attempt theorems 1.2.4–1.2.7 for locally compact groups there are many analytical complications; furthermore an argument valid for one of the means discussed in 2.2.1 may be utter nonsense for other types of means in that list. If we wish to prove one of these theorems for locally compact groups it is now clear, in view of 2.2.1, that it suffices to prove the theorem for any one of the means listed in 2.2.1. Previous efforts to extend these theorems can be unified and at the same time greatly simplified by examining the type of mean most appropriate to the task in each case. The generalized theorems apply to locally compact groups.

Theorem 2.3.1. If G is amenable and π a continuous homomorphism onto locally compact group H, then H is amenable.

Theorem 2.3.2. Every closed subgroup H of an amenable group G is amenable.

Theorem 2.3.3. If N is a closed normal subgroup in G and if N, G/N are amenable, then G is amenable.

Theorem 2.3.4. If G is a directed union of a system of closed amenable subgroups $\{H_\alpha\}$, in the sense that $G = \cup_\alpha H_\alpha$ and for any H_α, H_β there exists $H_\gamma \supset H_\alpha \cup H_\beta$, then G is amenable.

Proof (2.3.1): Consider LIM m on $CB(G)$. Then $f \in CB(H)$ $\Rightarrow f \circ \pi \in CB(G)$ and we can again take $\bar{m}(f) = m(f \circ \pi)$.

Q.E.D.

Proof (2.3.4): Consider means on $CB(G)$, then $f \in CB(G)$ $\Rightarrow f$ is a continuous, bounded function on each H_α and the discrete proof adapts verbatim. Q.E.D.

Proof (2.3.3): Consider $f \in UCB_r(G)$ and let m_1 be a LIM on $CB(N)$, m_2 a LIM on $CB(G/N)$. Then $x_j \to x$ in $G \Longrightarrow$

$$\|{}_{x_j}f - {}_xf\|_\infty \to 0 ;$$

hence $m_1({}_{x_j}f) \to m_1({}_xf)$ so that $F(x) = m_1({}_xf)$ is a continuous bounded function on G which is constant on cosets of N. Regarding F as a function in $CB(G/N)$ we may define:

$$\overline{m}(f) = m_2(F) = < m_2(x), m_1({}_xf) > \quad \text{all } f \in UCB_r(G) .$$

It is trivial that $\overline{m}$ is a LIM on $UCB_r(G)$. Q.E.D.

Proof (2.3.2): The problem here is to produce a transversal for the right cosets $G/H = \{Hx: x \in G\}$. If G is a separable (second countable) locally compact group we know there is a Borel measurable transversal (see introductory section of Mackey [51] or [16]) $T \subset G$ for right cosets of H with Borel measurable cross-section mapping $\tau: G/H \to G$. Let m be a LIM on $L^\infty(G)$ and $f \in CB(H)$. Extend f to Tf on G:

$$Tf(hx_a) = f(h) \qquad \text{all } h \in H, \ x_a \in T .$$

Since the cross-section map is Borel, Tf will be a Borel function on G and obviously $f = g$ in $CB(H)$ if $Tf = Tg$, a.e. on G, relative to the Haar measure in G. Thus $CB(H)$ is imbedded in $L^\infty(G)$ by a non-negative linear norm-decreasing map. We get the desired LIM on $CB(H)$ by defining:

$$\overline{m}(f) = m(Tf) .$$

Left invariance follows since $T({}_hf) = {}_h(Tf)$ for $h \in H$.

For a general locally compact G we shift our attention to left invariant means on $CB(G)$. First notice that

(*) if G is amenable and H an open subgroup, then H is amenable.

To prove (*), invoke the Axiom of Choice to produce a (discrete) transversal for the (open) cosets $\{Hx: x \in G\}$ and repeat the original argument for 1.2.5.

The general proof of 2.3.2 follows by considering an arbitrary (relatively open) σ-compact subgroup $H_\alpha \subset H$ and an open σ-compact subgroup $G_\alpha \subset G$ chosen so $H_\alpha \subset G_\alpha$. By a theorem of Kakutani-Kodaira [41], there is a compact normal subgroup $K_\alpha \subset G_\alpha$ such that G_α/K_α is separable. Now by (*) G_α is amenable, hence by 2.3.1 G_α/K_α is amenable. If π: $G_\alpha \to G_\alpha/K_\alpha$ is the canonical homomorphism, $\pi(H_\alpha) = \pi(H_\alpha K_\alpha)$ is a closed subgroup in G_α/K_α, so $\pi(H_\alpha)$ is amenable, and so is H_α since $\pi(H_\alpha) = H_\alpha K_\alpha/K_\alpha \cong H_\alpha/H_\alpha \cap K_\alpha$ and (2.3.3) applies to $e \to H_\alpha \cap K_\alpha \to H_\alpha \to \pi(H_\alpha) \to e$. But H is a directed union of its σ-compact open subgroups H_α and so H is amenable by 2.3.4. Q.E.D.

Remarks: By using the interplay between LIM on $L^\infty(G)$ and LIM on $CB(G)$ we have given a fairly elementary proof of 2.3.2, the most difficult of this family of results. Rickert [66], studying invariant means on $CB(G)$, proved 2.3.2 using the relatively simple Kakutani-Kodaira approximation theorem once it was established for separable groups. To prove 2.3.2 for the latter he needs powerful structure theorems about finite dimensional groups, etc; because his considerations deal solely with $CB(G)$ he is faced with the problem of finding *smooth* local transversals to the right cosets of H. But it is easyto show that such groups have *Borel* transversals, so the proof (for separable groups) is simple if we consider invariant means on $L^\infty(G)$.

§ 2.4. THE CELEBRATED METHOD OF DAY

In [8] Day introduced the notions of weak and strong invariance which have been fundamental in applications of invariant means to the theory of locally compact groups, harmonic analysis, and representation theory.

Definition 2.4.1. A net $\{\phi_j\} \subset P(G)$ is *weakly (strongly) convergent to left invariance* if $_x(\phi_j) - \phi_j \to 0$ weak* in $(L^\infty)^*$ (in $\|\cdot\|_1$-norm) for each $x \in G$. It is *weakly (strongly) convergent to topological left invariance* if $\phi * \phi_j - \phi_j \to 0$ weak* in $(L^\infty)^*$ (in $\|\cdot\|_1$-norm) for all $\phi \in P(G)$.

Of course there are right-handed versions of weak (strong) invariance but left invariance is the notion connected with left invariant means. The central results are:

Theorem 2.4.2. There is a net in $P(G)$ weakly convergent to (topological) left invariance $\Longleftrightarrow$ there is a net in $P(G)$ strongly convergent to (topological) left invariance.

Theorem 2.4.3. There is a net in $P(G)$ weakly convergent to (topological) left invariance $\Longleftrightarrow$ G is amenable.

These were proved for discrete groups in [8]; Hulanicki employed convergence to topological left invariance extensively in [37].

Namioka gives what is perhaps the simplest proof of 2.4.2 in [54].

Proof (2.4.2): We do the (more difficult) topological version. Implication ($<=$) is trivial. Conversely, for each $\phi \in P(G)$ take a copy of $L^1(G)$, form the locally convex product space $E = \Pi\{L^1(G): \phi \in P(G)\}$ with the product of norm topologies, and define the linear map $T: L^1(G) \to E$

$$Tf(\phi) = \phi * f - f \qquad \text{all } \phi \in P(G), \quad f \in L^1(G).$$

Now the weak topology on E coincides with the product of weak topologies (see [44], p. 160). Since $\phi * \phi_j - \phi_j \to 0$ weak* in $(L^\infty)^*$ for each $\phi \in P(G)$, zero lies in the weak closure S^- of $S = T(P(G)) \subset E$. Since E is locally convex and $T(P)$ is a convex set, the weak and strong closures of S coincide, so there is some net $\{\psi_i\} \subset P(G)$ such that $T(\psi_i) \to 0$ in E; that is to say: $\|\phi * \psi_i - \psi_i\|_1 \to 0$ for all $\phi \in P(G)$.

Q.E.D.

Proof (2.4.3): If $\{\phi_j\} \subset P(G)$ converges weakly to left invariance, $\{\phi_j\}$ lies within the weak* compact convex set Σ of all means on $L^\infty(G)$ and, taking a subnet, must be weak* convergent to some mean $\phi_j \to m$. But m is a LIM, and G is thus amenable, since:

$$m({}_xf) - m(f) \leftarrow \phi_j({}_xf) - \phi_j(f)$$

$$= \int [f(x^{-1}t) - f(t)]\,\phi_j(t)\,dt$$

$$= \int f(t)[{}_{x^{-1}}\phi_j(t) - \phi_j(t)]\,dt \to 0$$

for all $x \in G$. On the other hand, if m is any LIM on $L^\infty(G)$, the weak $*$ density of $P(G)$ in the set of all means on L^∞ insures we can find a weak $*$ convergent net $\{\phi_j\} \subset P(G)$ such that $\phi_j \to m$. Now if $f \in L^\infty$ and $x \in G$ we have $m({}_{x^{-1}}(f)) = m(f)$, which implies

$$\begin{aligned} < {}_x\phi_j - \phi_j, f > &= < {}_x\phi_j, f > - < \phi_j, f > \\ &= < \phi_j, {}_{x^{-1}}(f) > - < \phi_j, f > \\ &= < \phi_j, {}_{x^{-1}}(f) > - < m, {}_{x^{-1}}(f) > \\ &\qquad + < m, f > - < \phi_j, f > \to 0 \end{aligned}$$

so $\{\phi_j\}$ converges weakly to left invariance. The same arguments show that a net convergent to topological left invariance corresponds to a topological LIM on L^∞. Q.E.D.

The first part of this proof demonstrates a useful principle.

Corollary 2.4.4. If $\{\phi_i\} \subset P(G)$ converges weakly to (topological) left invariance, then any weak $*$ limit point of $\{\phi_j\}$ in the set Σ of all means on L^∞ is a (topological) left invariant mean on L^∞.

Nets convergent to left invariance may often be constructed explicitly and provide an alternative approach to the study of problems involving invariant means. Here is an example of such construction and an application to show non-uniqueness of left invariant means (recall discussion of 1.1.4).

Let $G = \mathrm{R}$ and consider the intervals $E_n = [-n, n]$, $n = 1, 2 \ldots$; then if $\varepsilon > 0$ and compact set K are fixed we can find an index $n(K, \varepsilon)$ such that

$$\frac{|xE_n \Delta E_n|}{|E_n|} < \varepsilon \quad \text{all } x \in K, \ |E| = \text{Haar measure}$$

if $n \geq n(K, \varepsilon)$. The sequence of normalized characteristic functions $\phi_n = \phi_{E_n}$ is strongly convergent to left invariance, as $\|_x(\phi_n) - \phi_n\|_1 = |xE_n \Delta E_n|/|E_n|$, so any weak $*$ limit point of $\{\phi_n\} \subset (L^\infty)^*$ is a LIM. But there are many such limit points: let $f \in L^\infty$ be chosen so f is real-valued, $\|f\|_\infty = 1$, and $\limsup\{< \phi_n, f >\} = +1$, $\liminf\{< \phi_n, f >\} = -1$ and choose subnets $\{\phi_{n(j)}: j \in S\}$ and $\{\phi_{n(p)}: p \in P\}$ such that $< \phi_{n(j)}, f > \to +1$, $< \phi_{n(p)}, f > \to -1$. Taking any weak $*$ limit points m^+, m^- of the respective nets we get left invariant means on L^∞ with $< m^+, f > = +1$, $< m^-, f > = -1$ so that $m^+ \neq m^-$. Any convex sum $\lambda m^+ + (1-\lambda)m^-$ with $0 \leq \lambda \leq 1$ is also a LIM, so we have a continuum of distinct invariant means on $L^\infty(G)$.

As Granirer shows in [26], the vector subspace of left invariant means on $B(G)$ is actually infinite dimensional if G is an infinite amenable group.

SECTION 3

DIVERSE APPLICATIONS OF INVARIANT MEANS

We wish to make these results accessible to nonexperts, so we will try, in discussing applications of amenability to problems in functional analysis, to display the connection of each separate application with amenability as clearly and directly as possible rather than constructing a sequence of implications: (amenability) $\Rightarrow (A_1) \Rightarrow \cdots \Rightarrow$ (amenability), which might hopelessly obscure the direct relations which exist. Besides, elementary proofs can now be given in most of our applications.

§3.1 Means on weakly almost periodic functions

Let G be a locally compact group. We say that a function $f \in CB(G)$ is *almost periodic* ($f \in AP(G)$) if its orbit under right translations $O(f) = \{R_x f: x \in G\}$ is relatively compact in $CB(G)$ with respect to the norm topology (relatively compact means the closure is compact in the topology indicated). As is well known, the orbit $\{R_x f: x \in G\}$ under right translates is relatively compact in $CB(G) \Leftrightarrow$ the orbit $\{L_x f: x \in G\}$ under left translates is relatively compact; furthermore $AP(G)$ is a norm closed two sided invariant subalgebra of $CB(G)$ and $AP(G) \subset UCB(G)$—see [34], section 18, for details.

If G happens to be amenable there are (usually) many left invariant means on the larger function spaces

$$CB(G) \supset UCB_r(G) \supset UCB(G) \supset AP(G)\,;$$

but for *any* locally compact group G, amenable or not, $AP(G)$ has a unique left invariant mean (which is actually two-sided invariant), so all left invariant means on the larger spaces must coincide when restricted to $AP(G)$. Thus while the function spaces listed in 2.2.1 give equivalent theories of left invariant means, one cannot take spaces too much smaller than $UCB(G)$ without losing this pleasant behavior. An interesting problem is to consider the space $W(G)$ of all weakly almost periodic functions in $CB(G)$: $f \in CB(G)$ is *weakly almost periodic* if the orbit $O(f) = \{R_x f\colon x \in G\}$ is relatively compact with respect to the weak topology in $CB(G)$. Again: the left orbit of f is relatively compact $<=>$ the right orbit is relatively compact, as is shown in [30]; furthermore $W(G)$ includes $AP(G)$ and $C_0(G)$, the functions vanishing at infinity, $W(G)$ is a closed two-sided invariant subspace in $CB(G)$, and all functions in $W(G)$ are two-sided uniformly continuous: $W(G) \subset UCB(G)$, see Eberlein [14], sections 10–16. Furthermore it is known that a left invariant mean on $W(G)$ is unique, if one exists at all, and that existence of a LIM on $W(G)$ implies a number of interesting ergodic properties for functions in $W(G)$ (see [23], particularly section 5.8).

We have the natural question: Is there always a LIM on $W(G)$ (even if G is not amenable)? The desirability of having a LIM on $W(G)$ is a major theme in [23] where this question is probed (section 5). A difficult fixed point theorem discovered by Ryll-Nardzewski [68] allowed him to resolve this question: in fact *every* locally compact group G admits a LIM on $W(G)$. The essential fixed point theorem (which we prove in Appendix 2) is:

Theorem 3.1.1. (Ryll-Nardzewski). Let E be a Banach space, G any (discrete) semi-group acting on some weakly compact convex set $K \subset E$. Assume G is distal relative to the norm topology in E (zero lies in the norm closure of the orbit $G(x-y) \Longleftrightarrow x = y$ in E), Then there is a fixed point for G in K.

Ryll-Nardzewski's original proof was probabilistic, using the Martingale Convergence theorem; later some deep results on convexity and Banach spaces were applied to give a geometric proof (see Burckel [5]). We give an elementary proof due to Kelley and Asplund-Namioka in Appendix 2.

Once this result has been established, let $f \in W(G)$ be given. Its orbit in $CB(G)$ under the (distal!) action $f \to L_x f$, $O(f, L) = \{L_x f\colon x \in G\}$ is relatively compact in the weak topology of $CB(G)$, hence the weakly closed convex hull $C(f, L)$ of this orbit is weakly compact.[3] The affine action $f \to {}_x f = L_{x^{-1}} f$ has at least one fixed point, by 3.1.1, say $\lambda(f) \in C(f, L)$—evidently $\lambda(f)$ is a constant function on G. Similar considerations apply to the affine action $f \to f_x = R_x f$ on $C(f, R)$ and give a constant function $\rho(f) \in C(f, R)$.

Mazur's Theorem says that the weak and norm closures coincide for any convex set in a Banach space (see [74], p. 120, Theorem 2). Thus we can find a net of (finite) convex sums

$$S_a(f) = \Sigma\{\lambda(a, x) L_x f\colon x \in G\} \in C(f, L)$$

(3) For any weakly compact set in a Banach space, its weakly closed convex hull is also weakly compact. See [13], V. 6.4 or [44], section 17 and particularly 17.12. In any locally convex vector space X the closed convex hull of any compact set is totally bounded, hence compact if X is complete—see Bourbaki [3], II.4.1.

such that $\|S_\alpha(f) - \lambda(f)\|_\infty \to 0$. Likewise we have finite convex sums of right translates $\{T_\beta(f)\}$ with $\|T_\beta(f) - \rho(f)\|_\infty \to 0$. Obviously $S_\alpha(\rho(f)) = T_\beta(\rho(f)) = \rho(f)$ and $S_\alpha(\lambda(f)) = T_\beta(\lambda(f)) = \lambda(f)$ since $\rho(f)$, $\lambda(f)$ are constant functions; also

$$S_\alpha(T_\beta f) = T_\beta(S_\alpha f) \text{ since } R_x L_x = L_x R_x \text{ for all } x \in G.$$

We conclude that $\rho(f) = \lambda(f)$, hence that $\rho(f)$ and $\lambda(f)$ are *unique* fixed points in $C(f, R)$, $C(f, L)$ respectively, from the limits

$$0 \leftarrow \|\lambda(f) - S_\alpha(f)\|_\infty \geq \|\lambda(f) - T_\beta(S_\alpha f)\|_\infty \text{ all } \beta\,;$$

$$0 \leftarrow \|\rho(f) - T_\beta(f)\|_\infty \geq \|\rho(f) - S_\alpha(T_\beta f)\|_\infty \text{ all } \alpha\,.$$

The map M: $f \to \lambda(f)$ is clearly homogeneous and G-invariant with $M(1) = 1$ and $M(f) \geq 0$ if $f \geq 0$; if we can show M is additive, then M is the desired invariant mean on $W(G)$. For this we follow an argument devised by von Neumann in his study of invariant means on $AP(G)$. If $A = \{x_i : i \in I\}$ is a finite indexed family of points in G, write

$$S(A, f) = \frac{1}{|I|} \Sigma \{L_{x_i}(f) : i \in I\}$$

$$T(A, f) = \frac{1}{|I|} \Sigma \{R_{x_i}(f) : i \in I\}$$

where $|I|$ is the cardinality of I. It is clear that $S(A, f) \in C(f, L)$, $T(A, f) \in C(f, R)$, and

$$\|S(A, f)\|_\infty \leq \|f\|_\infty \qquad \|T(A, f)\|_\infty \leq \|f\|_\infty$$

(*) $S(A, T(B, f)) = T(B, S(A, f))$

$$S(A, S(B, f)) = S(BA, f) \qquad T(A, T(B, f)) = T(AB, f)$$

where we define indexed family $AB = \{z(i,j) = a_i b_j : (i,j) \in I \times J\}$. For a fixed vector $f \in W(G)$ the considerations of the last paragraph show that we have a convex combination $S(f) = \Sigma\, \alpha(x) L_x(f)$ such that $\|S(f) - M(f)\|_\infty < \varepsilon$. By altering S slightly we get a finite indexed family of points in G (repetitions allowed!) A such that $\|S(A,f) - M(f)\|_\infty < \varepsilon$, and similarly for right translations.

We assert that

(**) $M(T(B,f)) = M(f)$ all $f \in W(G)$, all indexed families B.

In fact, if $\varepsilon > 0$, there is a family C such that

$$\|S(C,f) - M(f)\|_\infty < \varepsilon,$$

which readily implies, by (*), that

$$\|S(C, T(B,f)) - M(f)\|_\infty = \|T(B, [S(C,f) - M(f)])\|_\infty < \varepsilon .$$

Letting C vary we can make the convex sums $S(C, T(B,f)) \in C(T(B,f), L)$ converge to the constant function $M(f)$, so $M(f) = M(T(B,f))$ due to the uniqueness of fixed points shown above.

Now consider $f, g \in W(G)$ and let B be chosen so that $\|T(B,g) - M(g)\|_\infty < \varepsilon$. then for *any* family A' we have

$$\|T(A'B, g) - M(g)\|_\infty = \|T(A', T(B,g)) - M(g)\|_\infty$$

$$= \|T(A', [T(B,g) - M(g)])\|_\infty < \varepsilon .$$

From (**) we see that there is a finite indexed family A such that

$$\|T(A, T(B, f)) - M(f)\|_\infty < \varepsilon,$$

hence $\|T(AB, f) - M(f)\|_\infty < \varepsilon$. Taking $A' = A$, we see that

$$\|T(AB, f+g) - M(f) - M(g)\|_\infty < 2\varepsilon .$$

But the sums $T(AB, f+g) \in C(f+g, L)$, and the only fixed point they can converge to is $M(f+g)$, so $M(f+g) = M(f) + M(g)$.

Q.E.D.

Example. Consider the sets $\{E_T: T > 0\}$ in $G = \mathrm{R}$, $E_T = [-T, T]$. If $\varepsilon > 0$ and compact set $K \subset G$ are fixed, there is some $T(e, K)$ such that $T \geq T(\varepsilon, K) \Longrightarrow$

$$(*) \qquad \frac{|xE_T \Delta E_T|}{|E_T|} < \varepsilon \quad \text{all } x \in K,$$

where $|S|$ = Haar measure; the normalized characteristic functions $\phi_T = \phi_{E_T}$ converge strongly to left invariance and any weak $*$ limit point m of the net $\{\phi_T\} \subset (L^\infty)^*$ is a LIM on L^∞ (recall 2.4.4). But m restricted to $W(G)$ is the (unique) invariant mean m on $W(G)$: $m(f) = M(f)$ all $f \in W(G)$. If we regard $\{\phi_T\}$ as a net in the (weak $*$ compact convex) set of means on $W(G)$ this shows that M is the only weak $*$ limit point of $\{\phi_T\}$; since there is at least one such limit point, by weak $*$ compactness of the set of means, we see $\phi_T \to M$ (weak $*$), and obtain the formula:

$$M(f) = \lim_{T \to \infty} \langle \phi_T, f \rangle = \lim_{T \to \infty} \left(\frac{1}{2T} \int_{-T}^{T} f(t)dt \right)$$

for all $f \in W(G)$.

In section 3.6 we will see that amenability of G allows us to construct (compact) sets U with $0 < |U| < \infty$ corresponding to given $\varepsilon > 0$, K compact in G, for which relation (*) holds with U in place of E_T. If the pairs $J = \{(e, K)\}$ are made into a directed set in the obvious way and if for each $j \in J$ we select a corresponding set U_j, we get the following general formula for the invariant mean on $W(G)$:

$$M(f) = \text{Lim}\left\{\frac{1}{|U_j|} \int_{U_j} f(t)dt: \; j \in J\right\}$$

(see [15], section 1 and also Hewitt-Ross [34], 18.10-18.14). There does not seem to be any formula like this for non-amenable G, although M exists on $W(G)$.

§3.2 Reiter's work in harmonic analysis (Reiter's Condition)

We have seen the straightforward connection between existence of a (topological) LIM on $L^\infty(G)$ and existence of a net $\{\phi_j\} \subset P(G)$ strongly convergent to (topological) left invariance:

$$(*) \qquad \|_x\phi_j - \phi_j\|_1 \to 0 \quad \text{all } x \in G$$

($\|\phi * \phi_j - \phi_j\|_1 \to 0$ all $\phi \in P(G)$ in the topological situation). Reiter has made extensive use of a condition (P_1) on G which is formally stronger than existence of a net with (*). This condition has been instrumental in simplifying many results in harmonic analysis of abelian locally compact groups (see especially [59]; formula (ii′), p. 405 is precisely the condition (P_1)).

(P_1) If $\varepsilon > 0$ and a compact set $K \subset G$ are given, there is some $\phi \in P(G)$ such that

$$\|_x\phi - \phi\|_1 < \varepsilon \qquad \text{all } x \in K.$$

If we partially order the system of pairs $J = \{(K, \varepsilon)\}$ in the obvious way and take ϕ_j as in (P_1) for $j \in J$ then $\{\phi_j\}$ evidently converges strongly to left invariance and G is amenable. Property (P_1) was demonstrated for abelian locally compact groups in [59], pp. 404-405 using the Plancherel formula. Most applications invoke commutativity only to insure that (P_1) holds, thus it is interesting to discover that (P_1) is equivalent to amenability. This result, and the proof we give, are due to Hulanicki [37].

Theorem 3.2.1. A locally compact group G is amenable $\iff$ G has property (P_1).

Proof: For the non-trivial ($\Rightarrow$) we know amenability implies existence of a net $\{\phi_j\} \subset P(G)$ strongly convergent to topological left invariance (see 2.4). Let $\varepsilon > 0$ and compact set $K \subset G$ be given and let β be a fixed element in $P(G)$. By picking a small compact neighborhood E of the unit in G we can insure

$$(1) \qquad \begin{aligned} &\|\phi_E * \beta - \beta\|_1 < \varepsilon \\ &\|_x\beta - \beta\|_1 < \varepsilon \qquad \text{all } x \in E \end{aligned}$$

where $\phi_E \in P(G)$ is the normalized characteristic function of E. Now select $\{x_1, \ldots, x_N\} \subset G$ so $\cup_{k=1}^N x_k E \supset K$ and set $\psi_k = \phi_{x_k E} [= {}_{x_k}(\phi_E)]$ for $k = 1, 2, \ldots N$ (assume $x_1 = e$, the unit). Since $\{\phi_j\}$ converges strongly to topological left invariance we can find some element ϕ_j with

$$(2)\qquad \begin{aligned} &\|\psi_k * \phi_j - \phi_j\| < \varepsilon \qquad k = 1, 2, \dots, N \\ &\|\beta * \phi_j - \phi_j\| < \varepsilon \,. \end{aligned}$$

We assert that $\phi = \beta * \phi_j \in P(G)$ is the element we need in (P_1). We will show

$$\|_{x_i t}\phi - \phi\| < 6\varepsilon \quad \text{for } i = 1, 2, \dots, N \text{ and } t \in E.$$

In fact, by (1):

$$\|\phi_E * \phi - {}_t\phi\| \leq \|\phi_E * \phi - \phi\| + \|\phi - {}_t\phi\| < 2\varepsilon$$

for $t \in E$, which $\Longrightarrow$ if $t \in E$ and $i = 1, 2, \dots, N$:

$$\|\phi_{x_i E} * \phi - {}_{x_i t}\phi\| = \|_{x_i}(\phi_E * \phi) - {}_{x_i}({}_t\phi)\| < 2\varepsilon.$$

This $\Longrightarrow$

$$\begin{aligned} \|_{x_i t}\phi - \phi\| &\leq 2\varepsilon + \|\phi_{x_i E} * \phi - \phi\| \\ &= 2\varepsilon + \|\phi_{x_i E} * \beta * \phi_j - \beta * \phi_j\| \\ &\leq 2\varepsilon + \|\phi_{x_i E} * \beta * \phi_j - \phi_{x_i E} * \phi_j\| \\ &\qquad + \|\phi_{x_i E} * \phi_j - \phi_j\| + \|\phi_j - \beta * \phi_j\| \,. \end{aligned}$$

Now use (2) to get the desired inequality. Q.E.D.

Rieter applied (P_1) for locally compact abelian groups in a number of directions: [59] includes studies of spectral synthesis, interpolation theorems for Fourier transforms, and homomorphisms of ideals in group algebras. Later [58] he showed that many ergodic properties of these groups arise as consequences of (P_1); his proofs in this article extend almost verbatim to arbitrary amenable groups via 3.2.1 if one takes careful account of the modular function. We shall discuss the details of such ergodic theorems in section 3.6. Later he and others studied the implications of (P_1) for arbitrary groups, see [60], [61], [62], [63], [64] and Dieudonne [10]; these results are now interpreted as consequences of amenability.

In Dieudonne [10] some interest was raised in the condition (P_q) for $1 \leq q < \infty$ (particularly $q = 2$):

(P_q) Given $\varepsilon > 0$ and any compact set $K \subset G$, there is some $\phi \in L^q(G)$ with $\phi \geq 0$ and $\|\phi\|_q = 1$, such that $\|_x\phi - \phi\|_q < \varepsilon$ for all $x \in K$.

The property (P_∞) is not of interest since it holds for any group (take $\phi \equiv 1$). The inequality: $|\alpha - \beta|^t \leq |\alpha^t - \beta^t|$ for $\alpha, \beta \geq 0$ and $t \geq 1$ shows that $(P_1) \Rightarrow (P_q)$: for if $\phi \in P(G)$ has $\|_x\phi - \phi\|_1 < \varepsilon^q$ for all $x \in K$, and $\psi = \phi^{1/q}$, then $\psi \in L^q$ and

$$\|_x\psi - \psi\|_q = \left(\int |_x\phi^{1/q} - \phi^{1/q}|^q \, dt\right)^{1/q}$$

$$\leq (\|_x\phi - \phi\|_1)^{1/q} < \varepsilon .$$

Stegeman [69] gave the following elementary proof that $(P_1) \Leftrightarrow (P_q)$ for $1 \leq q < \infty$, based on the same inequality. First, if (P_q) holds, then so does (P_r) for $r \geq q$: let $\phi \in L^q$ be chosen so $\phi \geq 0$, $\|\phi\|_q = 1$, $\|_x\phi - \phi\|_q < \varepsilon^{r/q}$ for all $x \in K$, and take $\psi = \phi^{q/r} \in L^r$. The same considerations as above show $\|_x\psi - \psi\|_r < \varepsilon$ for all $x \in K$. Next observe that $(P_{2r}) \Rightarrow (P_r)$ for $1 \leq r < \infty$: for if $\phi \in L^{2r}$ with $\|_x\phi - \phi\|_{2r} < \frac{1}{2}\varepsilon$ for all $x \in K$, we may take $\psi = \phi^2 \in L^r$ and from Schwarz' inequality we see:

$$\begin{aligned}\|{}_x\psi - \psi\|_r &= \left(\int |{}_x\phi^2 - \phi^2|^r\,dt\right)^{1/r}\\ &\le \left[\left(\int |{}_x\phi + \phi|^{2r}\,dt\right)^{1/2}\left(\int |{}_x\phi - \phi|^{2r}\,dt\right)^{1/2}\right]^{1/r}\\ &\le 2\|\phi\|_{2r}\cdot \|{}_x\phi - \phi\|_{2r} < \varepsilon\ .\end{aligned}$$

These observations suffice to prove $(P_1) <\!\!=\!\!> (P_q)$ for all $1 \le q < \infty$.

We apply these remarks to the following convolution problem introduced in [10]: if $\mu \in M(G)$, the bounded regular Borel measures, the left convolution operation $\lambda_{\mu,p}$: $f(t) \to \mu * f(t) = \int f(s^{-1}t)\,d\mu(s)$ is a bounded linear operator in $L^p(G)$ with $\|\lambda_{\mu,p}\| \le \|\mu\|$. For $p = 1$, $p = \infty$ we actually have $\|\lambda_{\mu,p}\| = \|\mu\|$ for any locally compact G: if $p = 1$ one considers any approximate identity $\{\phi_j\} \subset P(G)$ for $L^1(G)$, so $\|\phi_j * f - f\|_1 \to 0$ and $\|f * \phi_j - f\|_1 \to 0$ for all $f \in L^1(G)$, and notice that

$$\phi_j \xrightarrow{(\sigma)} \delta_e$$

in $M(G)$ [(σ) indicates weak$*$ convergence in $M(G) = C_0(G)^*$]. Thus by separate weak$*$ continuity of convolution,

$$\mu * \phi_j \xrightarrow{(\sigma)} \mu * \delta_e = \mu \text{ in } M(G).$$

Thus we have

$$\begin{aligned}\|\mu\| \ge \|\lambda_{\mu,1}\| &\ge \sup\{\|\mu * \phi_j\| : j \in J\}\\ &\ge \limsup\{\|\mu * \phi_j\| : j \in J\} \ge \|\mu\|,\end{aligned}$$

so $\|\lambda_{\mu,1}\| = \|\mu\|$. For $p = \infty$ take any continuous bounded function f and note that $\mu * f(e) = \int f(t^{-1})\,d\mu(t) = \langle \mu, f^{\sim} \rangle$ and $\|\mu * f\|_\infty \ge |\mu * f(e)|$ ($\mu * f$ being continuous): evidently $\|\lambda_{\mu,\infty}\| = \|\mu\|$ all $\mu \in M(G)$. Also note that, for μ fixed, if

$\|\lambda_{\mu,p}\| = \|\mu\|$ for some p, $1 < p < \infty$, we have $\|\lambda_{\mu,p}\| = \|\mu\|$ for all $1 \leq p \leq \infty$, since the Riesz Convexity Theorem (see [13], VI. 10.8) insures that $\log \|\lambda_{\mu, 1/a}\|$ is a convex function of $a \in [0,1]$.

Theorem 3.2.2: If G is amenable then $\|\lambda_{\mu,p}\| = \|\mu\|$ for all $\mu \geq 0$ in $M(G)$.

Note. There are simple counter examples to $\|\lambda_{\mu,p}\| = \|\mu\|$ if μ is not required $\mu \geq 0$, and $1 < p < \infty$ (try G = cyclic group of 3 elements). Actually, G is amenable $<\!\!=\!\!>$ $\|\lambda_{\mu,p}\| = \|\mu\|$ for all $\mu \geq 0$. For discrete G this was proved in Day [9]; for non-discrete G this can be seen from Leptin [46], Theorem 1 by using the above remarks about the Riesz Convexity theorem

Proof: It suffices to look at normalized $\|\mu\| = 1$ and show $\|\lambda_{\mu,p}\| \geq 1$. Also we may assume μ has compact support, $\operatorname{supp}(\mu) = K$ (these measures are norm dense in $M(G)$). Now G satisfies (P_p) for any $1 \leq p < \infty$; let $f \in L^p(G)$ with $f \geq 0$ $\|f\|_p = 1$ be chosen to satisfy the requirements of (P_p) and let $g \in L^q(G)$, where $\frac{1}{p} + \frac{1}{q} = 1$. Then, using Fubini's Theorem:

$$|<\mu * f, g> - <f, g>|$$

$$= \left| \int \left[\int f(t^{-1}x)\, g(x) dx \right] d\mu(t) - \int \left[\int f(x)\, g(x) dx \right] d\mu(t) \right|$$

$$\leq \int \left[\int |f(t^{-1}x) - f(x)|\, |g(x)| dx \right] d\mu(t)$$

$$\leq \int_K \|{}_t f - f\|_p \cdot \|g\|_q d\mu(t) < \varepsilon \|g\|_q .$$

so $\|\mu * f - f\|_p \leq \varepsilon$, which $\Longrightarrow$ $\|\mu * f\|_p \geq \|f\|_p - \varepsilon = 1 - \varepsilon$ if $\varepsilon > 0$, and $\|\lambda_{\mu,p}\| \geq 1$. Q.E.D.

§3.3. THE FIXED POINT PROPERTY

We say that a locally compact group G has the *fixed point property* if, whenever G acts affinely on a compact convex set S in a locally convex space E, with the map $G \times S \to S$ continuous, there is a point $s_0 \in S$ fixed under the action of G. Thus, to each $x \in G$ we associate an *affine mapping* T_x on S [(4)] $[T_x(\lambda s_1 + (1-\lambda)s_2) = \lambda T_x(s_1) + (1-\lambda) T_x(s_2)$, all $0 \leq \lambda \leq 1]$ with $T_{xy} = T_x \circ T_y$ and $(x, s) \to T_x(s)$ jointly continuous. In particular each T_x: $S \to S$ is continuous. In [20] Furstenberg defined this property and proved that a connected semi-simple Lie group does not have the fixed point property unless it is compact. Day [6] gave some results connecting the fixed point property with existence of a LIM on $CB(G)$ (but his proof in [6], showing that: there is a LIM on $CB(G)$ if G has the fixed point property, is wrong, see his correction [7]). The relationship between amenability and the fixed point property was finally established by Rickert [66], who gave the following definitive result.

Theorem 3.3.1: A locally compact group G has the fixed point property $\Longleftrightarrow$ G has a LIM on $UCB_r(G)$.

---

(4) If T is a continuous linear operator on E which leaves S invariant, $T|S$ is a continuous and affine mapping on S; but it is not at all clear that every continuous, affine map $T: S \to S$ must arise in this manner.

It is evident now, by applying 2.2.1, that Day's assertion in [6] is in fact true. This is an excellent example of a situation where proper choice of the invariant mean to be examined leads to a successful resolution of the problem. There is no correct proof of 3.3.1 for invariant means on $CB(G)$ in the literature.

Proof: (3.3.1). For ($\Longrightarrow$) consider the set Σ of means on $X = UCB_r(G)$, a weak $*$ compact convex set in X^*. Define $T_x(m)$ so $< T_x m, f > = < m, {}_{x^{-1}}f >$ for $x \in G$, $m \in X^*$, $f \in X$. Then this defines an affine group action of G on Σ and the map $G \times \Sigma \to \Sigma$ is continuous since $x_j \to x$ in $G \Longrightarrow$

$$\| {}_{x_j}(f) - {}_x(f) \|_\infty \to 0$$

(this is *not* true in larger spaces of functions such as $X = CB(G)$). Hence there is a fixed point $m \in \Sigma$ and m is the desired LIM:

$$m({}_x f) = < T_x^{-1}(m), f > = m(f) \qquad \text{for all } x \in G .$$

Conversely, let m be any mean on X and let G act affinely, continuously on compact convex set S in E. A standard compactness argument shows that $x \to < f^*, T_x(s) >$ is a bounded right uniformly continuous function on G for any continuous linear functional $f^* \in E^*$ and $s \in S$. Fix $s \in S$, and give E^* the weak$*$ topology. As S is compact,

$$< f^*, T_m(s) > = \int < f^*, T_x(s) > \, dm(x) \quad \text{all } f^* \in E^*$$

(the integral standing for the action of m on the function $x \to < f^*, T_x(s) >$) defines a continuous linear functional

$T_m(s) \in E^{**}$. We show that this functional arises from some point $s_0 \in S$; then we will prove that s_0 is the desired fixed point under the action of G if we take m to be a LIM on X.

Now $T_\mu(s) \in E^{**}$ is well defined for any mean μ on X with μ a finite sum of point masses: $\mu = \Sigma_{i=1}^N \lambda_i \delta_{x_i}$ with $\lambda_i \geq 0$, $\Sigma_{i=1}^N \lambda_i = 1$. It is clear that

$$T_\mu(s) = \sum_{i=1}^{N} \lambda_i T_{x_i}(s) \in S .$$

We have seen that these measures are weak $*$ dense in the set Σ of all means on X (see section 1.1). If we take a net of these measures $\{\mu_j\}$ weak $*$ convergent to our mean m then $s_j = T_{\mu_j}(s) \in S$ for each index. Let $A(S)$ be all affine continuous functions on S, so $A(S) \supset \{f^* | S: f^* \in E^*\}$; let τ_A be the weakest topology on S making these functions continuous. Obviously the identity map $S \to (S, \tau_A)$ is continuous, so S is compact when equipped with the τ_A topology; thus we may take a subnet of $\{T_{\mu_j}(s)\}$, if necessary, to get

$$T_{\mu_j}(s) \xrightarrow{(\tau_A)} s_0 \quad \text{for some } s_0 \in S .$$

Of course we still have $\mu_j \to m$ weak $*$ in $\Sigma \subset X^*$, so

$$\begin{aligned} < f^*, s_0 > &\leftarrow < f^*, T_{\mu_j}(s) > \\ &= \int < f^*, T_g(s) > d\mu_j(g) \to \int < f^*, T_g(s) > dm(g) \\ &= < f^*, T_m(s) > \end{aligned}$$

for all affine continuous functions $f^* \in A(S)$, and in particular for all $f^* \in E^*$. Thus $T_m(s) = s_0 \in S$.

Now if $f^* \epsilon E^*$ and if we write $f^* \circ T_x(s) = \langle f^*, T_x(s) \rangle$, then $f^* \circ T_x$ is obviously in $A(S)$; thus $T_m(s)$ is a fixed poin if m is a LIM on X since:

$$\begin{aligned} \langle f^*, T_x(T_m(s)) \rangle &= \langle f^* \circ T_x, T_m(s) \rangle \\ &= \int \langle f^* \circ T_x, T_g(s) \rangle \, dm(g) \\ &= \int \langle f^*, T_{xg}(s) \rangle \, dm(g) \\ &= \int \langle f^*, T_g(s) \rangle \, dm(g) = \langle f^*, T_m(s) \rangle \end{aligned}$$

all $f^* \epsilon E^*$. Q.E.D

Rickert [66] has extended a theorem of Furstenberg [20] on semi-simple Lie groups with the fixed point property to lo cally compact groups which are *almost connected* (G/G_0 con pact where G_0 is the identity component); we present his proof below.

Theorem 3.3.2: If G is a semi-simple locally compact group which is almost connected, then G has the fixed point property $\Longleftrightarrow$ G is compact.

Remark: Here we define the *radical:* $rad(G)$ of a locally co pact group to be the largest solvable connected normal subgroup in G; Iwasawa [38] has shown that the radical always exists and is a unique closed subgroup, and that $rad(G) = rad(G_0)$, where G_0 is the identity component. We say G is *semi-simple* if there are no (non-trivial) solvable connected normal subgroups in G, or equivalently if $rad(G)$ is trivial. Evidently $G/rad(G)$ is semi-simple for any locally compact group; then if we apply 2.3.1 and 2.3.3 to the exact sequence of continuous homomorphisms

$$(e) \to rad(G) \to G \to G/rad(G) \to (e),$$

and recall that every solvable group such as $rad(G)$ is amenable, we immediately obtain:

Corollary 3.3.3: If G is almost connected then G has the fixed point property $\iff$ $G/rad(G)$ is compact.

Proof: (3.3.2). First we give a self-contained proof for connected Lie groups.

Lemma: If G is a connected semi-simple Lie group, then G has the fixed point property $\iff$ G is compact.

Proof: If Z = center, then G is compact $\iff$ G/Z is compact (see [35], p. 144). Let G have the fixed point property (i.e., G is amenable). Then $G^* = G/Z$ has the fixed point property (use 3.3.1 and 2.3.1), and has trivial center [if Z^* = center (G^*), then its inverse image under $\pi: G \to G/Z$ is closed, normal, and central modulo Z (hence solvable); by semi-simplicity of G, $\pi^{-1}(Z^*)$ must be discrete. Since it is also normal and G is connected $\pi^{-1}(Z^*)$ is central, so $\pi^{-1}(Z^*) = Z$]. Thus G^* has an Iwasawa decomposition $G^* = KS$ (see [38], p. 525) where K is any maximal compact subgroup, S a corresponding closed simply connected solvable subgroup, and every $g \in G^*$ has a unique continuous factorization $g = k \cdot s$. But G^* acts on the (compact) space of left cosets $G^*/S = \{xS: x \in G^*\} = \{kS: k \in K\}$, hence it acts affinely on the probability measures $\Sigma = \{\mu \in M(G^*/S): \mu \geq 0, \|\mu\| = 1\}$. These form a convex weak $*$ compact set in $M(G^*/S)$ since G^*/S is compact; also, $G^* \times \Sigma \to \Sigma$ is continuous, so there is a $\mu \in \Sigma$ invariant under the action of G^*. This happens $\iff$ modular function of S coincides with the modular function of G^* restricted to S (see [34], p. 203-207); but G^* is unimodular (Helgason [33], p. 366), while S is not unless S is trivial (see explicit computation of this function in Ch. 10, [33]). Thus G cannot have the fixed point property unless G is compact. Q.E.D.

If G is any almost connected semi-simple group with the fixed point property (i.e., amenable), so is the closed subgroup G_0. Let K be a compact normal subgroup of G_0 such that G_0/K is a Lie group; then G_0/K has the fixed point property and G semi-simple $\Longrightarrow$ G_0 semi-simple (since $rad(G) = rad(G_0)$). We assert that G_0/K is also semi-simple.

Lemma (Rickert): If G is a semi-simple locally compact group, K a compact normal subgroup, then G/K is semi-simple.

Proof: Let N be the identity component of the centralizer of K in G. Then we have $G = KN$ (Iwasawa [38], p. 514), and by standard isomorphism theorems we have topological isomorphism $G/K = KN/K \cong N/K \cap N$, so we show $N/K \cap N$ semi-simple. Let $\pi: N \to N/K \cap N$ be the canonical homomorphism, R the radical of $N/K \cap N$, and $R_1 = \pi^{-1}(R)$—a closed normal subgroup in N. Evidently $N \cap K$ is central in N, so R_1 is solvable; since $G = KN$ and $kn = nk$ all $k \in K$, $n \in N$, every subgroup $H \subset N$ which is normal in N is also normal in G. Take $H = R_1$; since G is semi-simple, R_1 is a *totally disconnected* normal subgroup and so is $\pi(R_1) = R$. Hence $R = (e)$ and $N/N \cap K$ is semi-simple. Q.E.D.

Applying the first lemma we see G_0/K is compact, hence G_0 is compact and so is the almost connected group G. Q.E.D.

In another paper [65] Rickert has derived from 3.3.2, 3.3.3 the following remarkable result, which will not be proved here.

Theorem 3.3.4: If G is any locally compact group which is almost connected, then G has the fixed point property (G is amenable) $<=>$ no free group on two generators appears as a closed subgroup in G.

As we have indicated in 1.2 it is unknown whether such a result is true for discrete groups; if so it should be possible to give the ultimate characterization of amenable groups.

Using the equivalent definitions of amenability, it is not hard to show:

Theorem 3.3.5: A locally compact group G is amenable $<=>$ it has a fixed point whenever it acts affinely on a compact convex set S in a locally convex space E, with the map $G \times \Sigma \to \Sigma$ separately continuous.

Proof: Obviously this fixed point property is formally stronger than the original one, and thus implies amenability. Conversely, if G is amenable there is a LIM m on $CB(G)$ [note transition from means on UCB_r to $CB(G)$] by 2.2.1; if $G \times \Sigma \to \Sigma$ is a separately continuous affine action, the functions $g \to < g(f), f^* >$ are continuous bounded on G for $f \in S$, $f^* \in A(S)$, the continuous affine functions on S, since S is compact in E. As in the proof of 3.3.1, if $s \in S$ there is a unique element $T_m(s) \in S$ with

$$< f^*, T_m(s) > = \int < f^*, T_x(s) > dm(x) \quad \text{all } f^* \in E^*$$

and $T_m(s)$ is a fixed point: $T_x(T_m(s)) = T_m(s)$ all $x \in G$.

Q.E.D.

§3.4. A CLASSIC APPLICATION TO REPRESENTATION THEORY.

Here we present a slight generalization of results due to Sz-Nagy [70]. Let G be a locally compact group, H a Hilbert space with continuous linear operators $\mathcal{B}(H)$, and let $T: G \to \mathcal{B}(H)$ be a representation of G (as invertible bicontinuous linear operators) which is (wo)-continuous: for each pair $x, y \in H$ the function $g \to (T_g x, y)$ is continuous.

Theorem 3.4.1: If G is amenable and if the representation T is uniformly bounded: $a_T = \sup\{\|T_g\|: g \in G\} < \infty$, then T is similar to a (wo)-continuous *unitary* representation U, in the sense that there is an invertible bicontinuous linear operator $A \in \mathcal{B}(H)$ with $T_g = A^{-1} U_g A$ for all $g \in G$.

Proof: First we construct a new inner product $[x, y]$ which makes H a Hilbert space and unitarizes the operators T_g (so $[T_g x, T_g y] = [x, y]$ for all $x, y \in H$ and $g \in G$). This will be easy if T is (so)-continuous: i.e., for each $x \in H$ the mapping $g \to T_g(x)$ is continuous from G into H. The following result, whose elementary and self-contained proof may be found in Glicksberg-de Leeuw [22] p. 143-144, shows that (wo)-continuity of T implies (so)-continuity (this is a well-known result for *unitary* representations in Hilbert space).

Theorem: If G is a locally compact group, any (wo)-continuous representation as bicontinuous linear operators on some Banach space is (so)-continuous.

Now let m be a right invariant mean for $CB(G)$. For $x, y \in H$ set $[x, y] = < m(g), (T_g x, T_g y) >$. Obviously $[x, y]$ is a conjugate linear form on H with $[x, y] = [y, x]^{-}$; furthermore if $|x| = [x, x]^{1/2}$ we have:

$$a_T^{-1} \|x\|^2 \leq \inf\{|T_g(x)|^2 : g \in G\} \leq < m, |T_g x|^2 >$$
$$= [x, x] \leq \sup\{|T_g x|^2 : g \in G\} \leq a_T^2 \|x\|^2,$$

so our new inner product is equivalent to the old one and is complete. As is well known, there must be invertible bicontinuous, self-adjoint linear operator $A \in \mathcal{B}(H)$ such that $[x, y] = (Ax, Ay)$; thus for $x \in H$, $g \in G$ we have:

$$\|AT_g A^{-1}(x)\|^2 = |T_g A^{-1}(x)|^2 = < m(s), \|T_{sg} A^{-1}(x)\|^2 >$$
$$= < m(s), \|T_s A^{-1}(x)\|^2 > = |A^{-1}x|^2 = \|x\|^2$$

by right invariance, so $U_g = AT_g A^{-1}$ is unitary. Q.E.D.

Remark. Let E be a Banach space with continuous linear operators $\mathcal{B}(E)$ and let $T: G \to \mathcal{B}(E)$ be a representation of a (discrete) group as invertible bicontinuous linear operators. If T is uniformly bounded, so $a_T = \sup\{\|T_g\| : g \in G\} < \infty$, and if $\|x\|$ is the norm in E, it is not hard to show that $|x| = \sup\{\|T_g(x)\| : g \in G\}$ is a new norm on E with $a_T \|x\| \geq |x| \geq a_T^{-1} \|x\|$; furthermore, $|T_g x| = |x|$ for all $x \in E$, $g \in G$ so T represents G as isometries with respect to the new (equivalent) norm. Amenability of G is not relevant to this construction. Dixmier [11] suggests that validity of 3.4.1 might be *equivalent* to amenability of G. This question does not seem to be resolved in the literature.

As an application of 3.4.1 (several others are given in[11], p. 221-223), let S be set and $\mathcal{P}$ a collection of subsets closed under finite unions, intersections, differences, and with $\emptyset$ and S in $\mathcal{P}$. Let H be a Hilbert space and let $\mathcal{P}$ be represented by by a uniformly bounded family $\{E_\sigma : \sigma \in \mathcal{P}\}$ of idempotents in $\mathcal{B}(H)$: thus $E_\sigma^2 = E_\sigma$ (but not necessarily $E_\sigma = E_\sigma^*$) and

(1) $\sup\{\|E_\sigma\| : \sigma \in \mathcal{P}\} < \infty$;

(2) $E_\emptyset = 0$, $E_S = I$ (identity), $E_\sigma \cdot E_\tau = E_{\sigma \cap \tau}$;

(3) $E_\sigma + E_\tau = E_{\sigma \cup \tau}$ if $\sigma, \tau \in \mathcal{P}$ and $\sigma \cap \tau = \emptyset$.

Then there is a family $\{F_\sigma\}$ of orthogonal projections on H ($F_\sigma^2 = F_\sigma = F_\sigma^*$) which is similar to $\{E_\sigma\}$. Consider the family $\{2E_\sigma - I\}$; obviously this is a uniformly bounded group of bicontinuous operators in $\mathcal{B}(\mathcal{H})$, since $(2E_\sigma - I)^2 = I$ and $(2E_\sigma - I)(2E_\tau - I) = (2E_\omega - I)$ if we take

$$\omega = (\sigma \cap \tau) \cup ((S \backslash \sigma) \cap (S \backslash \tau)) .$$

As this is an *abelian* group, there is a similarity transformation $A \in \mathcal{B}(H)$ such that $\{A(2E_\sigma - I)A^{-1}\}$ is a family of unitary operators, so that $\{AE_\sigma A^{-1}\}$ are orthogonal projections. This equivalence can be useful in developing an operational calculus for operators on Banach spaces—see Lorch [48], [49], [50].

§3.5. WEAK CONTAINMENT OF IRREDUCIBLE REPRESENTATIONS

The left regular representation L of a locally compact group G is simply the action $L_x: f \to {}_xf$ on functions in $L^2(G)$. If U is any unitary representation of G on Hilbert space $H(U)$ it is well known that there is a one-one correspondence between such representations of G and (bounded) $*$-representations U of $L^1(G)$ which are nowhere trivial in the sense that $(0) = \bigcap \{\text{Ker}(U_f): f \in L^1(G)\}$. This correspondence is effected by taking

$$(U_f(\xi), \eta) = \int_G (U_x(\xi), \eta) \quad f(x)dx$$

for $f \in L^1(G)$ and $\xi, \eta \in H(U)$.

Many authors have been concerned with the natural questions:

(1) Which irreducible unitary representations of G arise within the regular representation?

(2) For which groups is every irreducible unitary representation contained in the left regular representation?

One way of defining "containment" is to consider direct integral decompositions; however there is an (equivalent) notion which avoids the complexities of direct integral theory.

Definition 3.5.1: Let U, V be unitary representations of G on Hilbert spaces $H(U), H(V)$, let U', V' be their extensions to $*$-homomorphisms of $L^1(G)$, and let E_U, E_V be the C^* algebras spanned by the ranges of U', V'. Then U *weakly contains* V if there is a $*$-homomorphism π mapping E_U onto E_V which makes the following diagram commute.

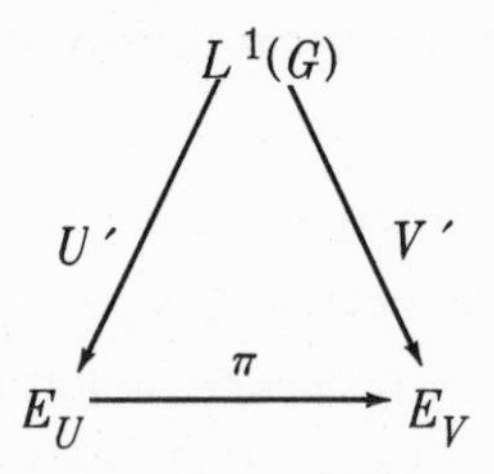

Note: Any $*$-homomorphism of C^* algebras such as π must be bounded (in fact $\|\pi\| \leq 1$). Weak containment in this sense is equivalent to requiring that any positive definite function associated with V be a uniform on compacta limit of linear combinations with positive coefficients of the positive definite functions associated with U; this is the more common definition of weak containment due to Fell [17], [18] (see also Dixmier [12], 58-82).

Fell [18] p. 391-401 has shown that the irreducible representation of $SL(n, C)$ weakly contained in the regular representation consist precisely of the principal non-degenerate series described by Gelfand-Naimark. He also showed (see [18], section 6) that the irreducible representations weakly contained in a given representation U are essentially those appearing when U is decomposed as a direct integral. Yoshizawa [73] proved that the free group on two generators does not have the weak containment property (2), and Takenouchi [71] proved that an almost connected locally compact group G (G/G_0 compact) has property (2) $<=>$ it is a (C)-group as defined in Iwasawa [38], (an almost connected group G is a (C) group $<=>$ $G/rad(G)$ is compact–see Rickert [65], section 5). Later Hulanicki [36], considering discrete groups, proved that G has the weak containment property $<=>$ there is a LIM on $B(G)$, however his proof of $(=>)$ is not correct as given; although the gaps can be repaired in this proof, it has been superseded by his more elegant proof in [37] where it is shown:

Theorem 3.5.2: A locally compact group G has the weak containment property $<=>$ G is amenable.

In the following proof this relation is shown more directly than it is in [37]; furthermore, we avoid the intermediate use of Følner conditions on G which in themselves are fairly deep consequences of amenability.

First there are a few preliminary simplifications. In terms of positive definite functions on G the weak containment property means we must be able to find, for any continuous positive definite f, a net of functions $\{\psi_a\}$ consisting of finite linear combinations $\psi_a = \Sigma_i \phi_i * \phi_i^{\sim}$ with $\phi_i \in L^2(G)$ which is uniformly on compacta convergent to f (see Godement [24]). In [24] Godement has also shown that all irreducibles are weakly contained in the regular representation L if and only if the one-dimensional identity representation is weakly contained in L, and that this happens $<=>$ there is a net $\{\phi_j\}$ in $L^2(G)$ such that $\phi_j * \phi_j^{\sim} \to 1$ uniformly on compacta in G. This latter property (equivalent to weak containment) will be shown to hold $<=>$ G is amenable.

Proof 3.5.2: If G is amenable, Reiter's condition (P_1) holds and if $\varepsilon > 0$ and compact set $K \subset G$ are given, there is some $\phi \in P(G)$ such that $\|_x\phi - \phi\|_1 < \varepsilon^2$ for all $x \in K$. Now consider $\psi = \phi^{1/2}$, a non-negative function in $L^2(G)$ with $\|\psi\|_2 = 1$. The well-known inequality $|\alpha-\beta|^2 \leq |\alpha^2-\beta^2|$ for

$\alpha, \beta \geq 0$ implies:

$$\|{}_x\psi - \psi\|_2 = \left[\int |\phi^{1/2}(x^{-1}t) - \phi^{1/2}(t)|^2 dt\right]^{1/2}$$
$$\leq \left[\int |\phi(x^{-1}t) - \phi(t)| dt\right]^{1/2} < \varepsilon$$

for all $x \in K$, hence $\int \psi(t)^2 dt = 1$ and (recall $\psi^\sim(x) = \psi(x^{-1})$ here)

$$|1 - \psi * \psi^\sim(x)| = |1 - \int \psi(t)\, \psi(x^{-1}\, t) dt|$$

$$= |\int \psi(t)\, \psi(t) dt - \int \psi(t)\, \psi(x^{-1}t) dt|$$

$$= |(\psi, \psi - {}_x\psi)| \leq \|\psi\|_2 \cdot \|{}_x\psi - \psi\|_2 < \varepsilon$$

for $x \in K$ as desired.

For the converse we follow [37]: let G have the weak containment property. If $\varepsilon > 0$ and compact set $K \subset G$ are specified, there is some $\psi \in L^2(G)$ such that $|\psi * \psi^\sim(x) - 1| < \varepsilon$ all $x \in K$. We may assume the unit in G is in K. By replacing ψ with $\psi' = a^{-1/2} \cdot \psi$, where $a = \|\psi\|_2 = \psi * \psi^\sim(e)$ [note that $|1 - a| < \varepsilon$ by our hypothesis], we can assume $1 = \psi * \psi^\sim(e) = \|\psi\|_2$. Let $f(t) = |\psi(t)|$, then

$$0 \leq |\psi * \psi^\sim(t)| = |\int \psi(s)\, \psi(t^{-1}s) ds| \leq f * f^\sim(t) \text{ all } t \in G$$

which $\Longrightarrow$

$$0 \leq 1 - f * f^\sim(t) \leq 1 - |\psi * \psi^\sim(t)| \leq |1 - \psi * \psi^\sim(t)| < \varepsilon$$

for all $x \in K$, which $\Longrightarrow$

$$|1 - (f, {}_xf)| = |1 - \int f(t)\, {}_xf(t) dt| = |1 - f * f^\sim(x)| < \varepsilon$$

for all $x \in K$, which $\Longrightarrow$

$$\|f - {}_x f\|_2^2 = (1-(f, {}_x f)) + (1-(f, {}_x f)) < 2\varepsilon \quad .$$

Now set $\phi = f^2$ so $\phi \in P(G)$. For all $x \in K$

$$\begin{aligned} \|\phi - {}_x\phi\|_1 &= \int |f^2(t) - {}_x f^2(t)|dt \\ &= \int |f(t) - {}_x f(t)|\ |f(t) + {}_x f(t)|dt \\ &\leq 2\|f - {}_x f\|_2^2 < 4\varepsilon , \end{aligned}$$

hence there is a net in $P(G)$ strongly convergent to left invariance and G is amenable. Q.E.D.

Another question related to the weak containment property concerns the *group C^*-algebra:* $C^*(G)$. The left regular representation L, when extended to a $*$-representation of $L^1(G)$ as bounded operators on $L^2(G)$ provides a natural C^*-norm (not necessarily complete of course) on $L^1(G)$. There is also the *greatest C^*-norm* on $L^1(G)$, $|f| = \sup\{\|T_f\|\}$ where the sup is taken over all nowhere trivial $*$-representations T of $L^1(G)$ as an algebra of bounded operators on Hilbert space. From the point of view of representation theory the completion $C^*(G)$ of $L^1(G)$ relative to $|f|$ is the natural C^*-algebra associated with G; however the operator norm $\|L_f\|$ on $L^2(G)$ is just $\|L_f\| = \sup\{\|f * \psi\|_2 \colon \|\psi\|_2 = 1\}$, and gives the C^*-algebra norm which is easiest to compute. The natural question is: when is $|f| = \|L_f\|$ for all $f \in L^1(G)$? It is known that this is equivalent to the weak containment property, hence equivalent to amenability of G. We *always* have $|f| \geq \|L_f\|$ if $f \in L^1(G)$.

§3.6. FØLNER'S CONDITION.

For discrete groups Følner [19] proved, via ingenious combinatory arguments, the fundamental result:

Theorem 3.6.1: There is a LIM on $B(G)$ $\Longleftrightarrow$ G satisfies the following condition

(FC) Given $\varepsilon > 0$ and any finite set $K \subset G$ there is a finite non-empty set $U \subset G$ with

$$\frac{1}{|U|} \; |xU \; \Delta \; U| < \epsilon \text{ for all } x \in K$$

where $|U|$ is the cardinality of U.

One can go further [54] and show that if a second finite set $E \subset G$ is specified in addition to K one can find U such that (FC) holds and $U \supset E$, which shows that the finite sets U satisfying (FC) relative to fixed (ε, K) can be chosen arbitrarily large.

It is now known that the topological analog of the theorem is valid.

Theorem 3.6.2: For any locally compact group, G is amenable $\Longleftrightarrow$ G has the property

(FC) Given $\varepsilon > 0$ and compact set $K \subset G$ there is a Borel set $U \subset G$ with $0 < |U| < \infty$ and

$$\frac{1}{|U|} \cdot |xU \; \Delta \; U| < \varepsilon \text{ for all } x \in K$$

where $|U|$ is left Haar measure.

Note. If an additional compact set $E \subset G$ is specified we can arrange that (FC) holds and $U \supset E$; we will not prove this—see [15], section 2. Regularity of Haar measure insures that we can take U a compact Baire set if we wish. Intuitively property (FC) says we can find compact sets $U \subset G$ which are "ε-large" relative to any fixed compact set $K \subset G$ of left translations.

It is trivial that (FC) $\Longrightarrow$ G is amenable, for if $J = \{(\varepsilon, K)\}$ is partially ordered in the obvious way, and for each $j \in J$ we pick an appropriate compact set U_j which is ε-large relative to K, then the net $\{\phi_j\}$ of normalized characteristic functions has

$$\|{}_x\phi_j - \phi_j\|_1 = \frac{1}{|U_j|} \, |xU_j \,\Delta\, U_j| \to 0$$

for each $x \in G$ so $\{\phi_j\} \subset P(G)$ is strongly convergent to left invariance and G is amenable.

Namioka [54] has given an elegant functional analytic proof of Følner's Theorem 3.6.1 which can be modified to prove that the following, measure-theoretic, *weak Følner condition* holds if G is amenable.

(FC*) Given $\varepsilon > 0$, $\delta > 0$ and any compact set $K \subset G$ there exist Borel sets $U \subset G$ and $N \subset K$ such that $0 < |U| < \infty$, $|N| < \delta$, and

$$|xU \,\Delta\, U|/|U| < \varepsilon \text{ for all } x \in K \backslash N.$$

The proof of (FC*) gives no indication how badly the exceptional set N is distributed in K, nor does it give us any idea how U is located in G, or how it behaves as we let $\delta \to 0$; hence (FC) is not immediately accessible by this sort of analysis. The ($\Longleftarrow$) part of 3.6.2 was proved for a large class of locally compact groups in Greenleaf [28] by methods which effectively show how the set U may be constructed (we will return to this largely unexplored localization problem at the end of this section). C. Ryll-Nardzewski has independently devised an elegant proof[5] that (FC*) => (FC) which we pre sent below. Thus the converse part of (3.6.2) is immediate from the following results.

Theorem 3.6.3: Amenability of a locally compact group $G \Rightarrow$ G has property (FC*).

Proof (Namioka): Let $\varepsilon > 0$ and $K \subset G$ any compact set wit $0 < |K| < \infty$ (if $|K| = 0$ there is nothing to prove, since we may take $N = K$).

Amenability of G implies that Reiter's condition (P_1) holds fc G: there is a $\phi \in P(G)$ such that $\|{}_x\phi - \phi\|_1 < \delta\varepsilon/|K|$ for a $x \in K$. We may take ϕ to be a simple function if we wish anc so may write $\phi = \Sigma_{i=1}^N \lambda_i \phi_i$ where $A_1 \supset \cdots \supset A_N$ are Borel sets with $0 < |A_i| < \infty$, ϕ_i are the normalized characteristic functions $\phi_i = \phi_{A_i} \in P(G)$, and where $\lambda_i > 0$, $\Sigma_{i=1}^N \lambda_i = 1$. As indicated in [54] it is a straightforward matter to show:

$$\|{}_x\phi - \phi\|_1 = \sum_{i=1}^{N} \lambda_i \frac{|xA_i \,\Delta\, A_i|}{|A_i|} < \frac{\varepsilon\delta}{|K|}$$

since $A_1 \supset \cdots \supset A_N$. Integrating over $x \in K$ we get

$$\delta\varepsilon > \sum_{i=1}^{N} \lambda_i \int_K |xA_i \,\Delta\, A_i| / |A_i| \, dx$$

---

(5) This proof was never announced. I am indebted to I. Namioka and A. Hulanicki for communicating the proof to me.

and since this is a convex sum we must have

$$\int_K |xA \,\Delta\, A| / |A| dx < \varepsilon\delta$$

for at least one $A = A_i$. The integrand can be $\geq \varepsilon$ only on a set $N \subset K$ with $|N| < \delta$, so we have $|xA \,\Delta\, A| / |A| < \varepsilon$ for $x \in K \setminus N$. Q.E.D.

Lemma 3.6.4. For any locally compact group (FC*) $\Rightarrow$ (FC).

Proof (Ryll-Nardzewski, unpublished): Let $K \subset G$ be compact with left Haar measure $|K| > 0$ and let $A = K \cup KK$; then for each $k \in K$, $|kA \cap A| \geq |kK| = |K|$. Let $\delta = \frac{1}{2}|K|$, then for any subset $N \subset A$ such that $|A \setminus N| < \delta$, and for any $k \in K$ we have

$$2\delta \leq |K| \leq |kA \cap A| \leq |kN \cap N| + |A \setminus N| + |k(A \setminus N)|$$

$$< |kN \cap N| + 2\delta\ .$$

Hence $|N \cap kN| > 0$, which implies $k \in NN^{-1}$ (or equivalently $K \subset NN^{-1}$).

Now apply (FC*) to $\varepsilon/2 > 0$, δ as above, and compact set A. Then there is a (compact) set U with $0 < |U| < \infty$, and $N \subset A$ such that $|A \setminus N| < \delta$, such that $|nU \,\Delta\, U| < \frac{\varepsilon}{2}|U|$ for all $n \in N$. For $n_1, n_2 \in N$ this means:

$$|n_1 n_2^{-1} U \,\Delta\, U| \leq |n_2^{-1} U \,\Delta\, U| + |U \,\Delta\, n_1^{-1} U|$$

$$= |n_2 U \,\Delta\, U| + |n_1 U \,\Delta\, U| < \varepsilon |U|,$$

so that $|xU \,\Delta\, U| < \varepsilon|U|$ for all $x \in K \subset NN^{-1}$, proving (FC). Q.E.D.

Note: The argument for 3.6.3 can be modified to prove that G has property

(FC**) If $\varepsilon > 0$ and K is any *finite* set in G, then there is a Borel set $U \subset G$ with $0 < |U| < \infty$ and

$$|xU \,\Delta\, U| \,/\, |U| < \varepsilon \quad \text{for all } x \in K.$$

This is mentioned explicitly in [54]. Hulanicki [37] gives a complicated measure theoretic Følner condition which subsumes both (FC*) and (FC**) but is weaker than (FC).

As we have indicated, in an amenable group G with (ε, K) given, it is one thing to prove the existence of a set U which is ε-large relative to K, and quite another thing to give an effective procedure for constructing U. For example one sees easily that the sets $\{E_n = [-n, n]\colon\ n = 1, 2, \ldots\}$ in $G = \mathrm{R}$ are eventually ε-large relative to any fixed compact set; however consider the "$ax + b$" group which is $G = \{(a, x)\colon\ a, x \in \mathrm{R}\}$ equipped with product operation: $(a, x)(b, y) = (a + e^x b, x + y)$, which makes G the semi-direct product of closed normal subgroup $N = \{(a, 0)\colon\ a \in \mathrm{R}\}$ and closed (non-normal) subgroup $H = \{(0, x);\ x \in \mathrm{R}\}$. One can easily carry out the computations necessary to show that *rectangular sets* like $A \times X = \{(a, x)\colon\ a \in A, x \in X\}$ can never be ε-large for the one point set $K = \{(0, -1)\}$ for $0 < \varepsilon < \frac{1}{2}$, no matter how cleverly we select A, X in N, H; in some sense rectangular sets have the wrong "shape" to be large relative to compacta in G. From the considerations given in [28] one can effectively construct a large U for a prescribed pair (ε, K), since we know how to do this in each of the component groups N, H: let K_H, K_N be the obvious projections onto H, N; let compact set $X \subset H$ be $\frac{\varepsilon}{2}$-large relative to $K_H \subset H$ and $A \subset H$ $\frac{\varepsilon}{2}$-large relative to the compact set $X^{-1}K_N \subset N$; then

$$U(\varepsilon, K) = X \cdot A = \{(0, x)(a, 0) = (e^x a, x): \ a \in A, \ x \in X\}$$

is the desired set in G which is ε-large relative to K. Similar constructions work for semi-direct products and group extensions.

Finally, there is the following *localization conjecture* which is by and large unproved.

Conjecture. Let G be a (connected, separable) amenable locally compact group and U any compact symmetric neighborhood of the unit. Then $G = \bigcup_{n=1}^{\infty} U^n$ is well known; is it true that the sets $\{U^n: n = 1, 2, \ldots\}$ are *eventually* ε-large relative to any fixed compact set of left translations in G?

Kawada [43] stated the following theorem, but his proof has several serious gaps; these are corrected in [76]. The arguments use many special properties of vector groups.

Theorem 3.6.5. If U is a compact neighborhood of the unit in a connected locally compact abelian group, then the sets $\{U^n\}$ are eventually large with respect to (ε, K) for any compact set $K \subset G$.

The growth of sets $\{U^p\}$ where U is a finite set in a discrete abelian group is discussed in [76]. Here is another suggestive localization theorem.

Theorem 3.6.6. Let G be any σ-compact locally compact abelian group and U any compact neighborhood of the unit such that

$$G = \bigcup_{n=1}^{\infty} U^n .$$

If $\varepsilon > 0$ and compact set $K \subset G$ are given, *infinitely many* of the sets $\{U^n\}$ are ε-large relative to K.

Proof. Let $V = U^{p_0}$ be taken so $V \supset KK^{-1}$ and choose a symmetric set of points $X\ (=X^{-1}) = \{x_1 = e, x_2, \ldots, x_N\} \subset V$ such that $V^2 \subset \cup\{x_i V : i = 1, 2, \ldots, N\}$, which $\Longrightarrow$

$$\begin{aligned} V^{p+1} &\subset \cup\{x_{i_1} \cdots x_{i_p} V : 1 \leq i_k \leq N\} \\ &= \cup\{x_1^{\lambda_1} \cdots x_N^{\lambda_N} V : \sum_{i=1}^{N} \lambda_k = p,\ \lambda_k \text{ integers} \geq 0\}; \end{aligned}$$

the latter union has at most p^N distinct sets in it, so that $|V^{p+1}| \leq p^N |V|$ where $|E|$ = Haar measure of E. This insures $\liminf\{|V^{p+1}| / |V^p|\} = 1$, for otherwise there is some $\delta > 0$ such that $|V^{p+1}| \geq (1+\delta)|V^p|$ for all large p, which $\Longrightarrow$ $|V^p|$ has *exponential* growth in p. Now $x \in KK^{-1} \Longrightarrow$ $x^{-1} \in KK^{-1}$ and obviously $\{xV^p, x^{-1}V^p\} \subset V^{p+1}$; since $|xV^p \backslash V^p| = |V^p \backslash x^{-1}V^p|$ we see:

$$0 \leq \frac{|xV^p \,\Delta\, V^p|}{|V^p|} \leq 2\,\frac{|V^{p+1} \backslash V^p|}{|V^p|} = 2\left(\frac{|V^{p+1}|}{|V^p|} - 1\right)$$

The right side has $\liminf = 0$, independent of $x \in KK^{-1}$, so the sets $\{V^p\}$ are frequently ε-large relative to KK^{-1} (hence also to K). Q.E.D.

A similar estimate on the size of $\{x_{i_1} \cdots x_{i_p} : 1 \leq i_k \leq N\}$ as $p \to \infty$ proves 3.6.6 for nilpotent G, but solvable groups do not yield to this line of attack. For further comments see [28], [54], and Hewitt-Ross [34], 18.10-18.14.

Quite recently Leptin [45], [46] has shown that, for the same class of locally compact groups studied in [28], G amenable implies

(A) If $\varepsilon > 0$ and $K \subset G$ is a compact set including the unit, there is a compact set U with $0 < |U| < \infty$ and
$$\frac{1}{|U|}\,|KU \,\Delta\, U| < \varepsilon .$$

Note that (A) => (FC), for if U satisfies (A) with respect to $(\frac{\varepsilon}{2}, KK^{-1})$ then

$$|kU \,\Delta\, U| = |kU \setminus U| + |U \setminus kU| = |kU \setminus U| + |k^{-1}U \setminus U|$$
$$\leq 2\,|KK^{-1}U \setminus U| < \varepsilon |U| \quad \text{for all } k \in K .$$

For discrete groups (FC) => (A): given $K = \{x_1, \ldots, x_n\} \supset \{e\}$ pick U satisfying (FC) for $(\varepsilon/N, KK^{-1})$. Then

$$0 \leq |KU \,\Delta\, U| = |\bigcup_{i=1}^{N} (x_i U \,\Delta\, U)| \leq \sum_{i=1}^{N} |x_i U \,\Delta\, U| < \frac{\varepsilon}{N} \cdot N|U|$$
$$= \varepsilon |U| .$$

Leptin [45] was interested in the group invariant ($\mathcal{K}$ = compacta in G):

$$I(G) = \sup\left\{\inf\left\{ \frac{|KU|}{|U|} : U \in \mathcal{K},\ |U| > 0\right\} : K \in \mathcal{K}\right\} \geq 1 .$$

It is easily seen that $I(G) = 1$ <=> G satisfies (A). The result: (amenability) <=> (A) is valid for all locally compact groups, see [15] for details. It is interesting to note that for all groups which have property (A) one may resolve affirmatively the "L^p-conjecture" discussed by Rajagopalan [57] and Zelazko [75].

Conjecture. If $1 < p < \infty$ and if G is a locally compact group, then $L^p(G)$ closed under convolution ($f, g \in L^p \Longrightarrow f * g \in L^p$) implies G is compact.

The category of groups for which this conjecture is true is not related to the category of amenable groups (it seems as if it should be true for *all* groups); the above remarks show that all amenable groups are known to belong to this category.

§3.7 ERGODIC PROPERTIES OF AMENABLE GROUPS

Let G be a locally compact group with left Haar measure dt and let $f \in L^1(G, dt)$. Consider the associated measure $\mu_f \in M(G)$ such that $d\mu_f(t) = f(t)dt$ and the convex hull $C_r(f) \subset M(G)$ of its right translates

$$\left\{ \sum_{i=1}^{N} \lambda_i \mu_f * \delta_{x_i} : \ \lambda_i \geq 0, \ \Sigma_i \lambda_i = 1, \ x_i \in G \right\}.$$

Taking the modular function Δ on G into account one can verify

$$\left\| \sum_{i=1}^{N} \lambda_i \mu_f * \delta_{x_i} \right\| = \int_G \left| \sum_{i=1}^{N} \lambda_i f(t x_i^{-1}) \ \Delta(x_i^{-1}) \right| dt .$$

For left translates $C_\ell(f) = \{\Sigma_i \lambda_i \delta_{x_i} * \mu_f\}$ the modular function does not appear and:

$$\left\| \sum_{i=1}^{N} \lambda_i \delta_{x_i} * \mu_f \right\| = \int_G \left| \sum_{i=1}^{N} \lambda_i f(x_i^{-1} t) \right| dt .$$

Reiter studied groups satisfying condition (P_1) in [58] and proved they have a number of ergodic properties, among which the following is typical.

Theorem 3.7.1. Let G be any locally compact group satisfying property (P_1). Then the distance from the origin to the convex hull $C_r(f)$:

$$d(0, C_r(f)) = \inf\{\| \Sigma_i \lambda_i \mu_f * \delta_{x_i} \|\}$$

is precisely the modulus of the average value of f

$$d(0, C_r(f)) = |\int f(t)dt| = |< \mu_f, 1 >| .$$

This is a consequence of the following more general result. Let G be a locally compact group, $H \subset G$ a closed subgroup (not necessarily normal), and consider the right and left coset spaces $(G/H)_r = \{Hx: x \in G\}$, $(G/H)_\ell = \{xH: x \in G\}$ with their usual topologies and canonical mappings ρ, λ respective, ly. There is a natural norm decreasing *linear* mapping ρ^{**}: $M(G) \to M((G/H)_r)$ which "averages" $\mu \in M(G)$ over right cosets of H:

$$< \rho^{**}(\mu), \psi > = < \mu, \psi \circ \rho > \text{ for } \psi \in C_0((G/H)_r) .$$

Notice that $\psi \circ \rho$ is continuous, bounded, and constant on right cosets of H, hence is μ-integrable. Similarly define λ^{**}: $M(G) \to M((G/H)_\ell)$, and consider the action of H on $L^1(G, dt)$ by right, left translation.

Theorem 3.7.2. Let G be locally compact, H a closed subgroup which has property (P_1), and let $f \in L^1(G, dt)$. Then

$$\inf\left\{\| \sum_{i=1}^{N} \lambda_i \mu_f * \delta_{x_i} \| : \sum_{i=1}^{N} \lambda_i = 1 \text{ with } \lambda_i \geq 0,\ x_i \in H\right\} = \|\lambda^{**}(\mu_f)\|$$

$$\inf\left\{\| \sum_{i=1}^{N} \lambda_i \delta_{x_i} * \mu_f \| : \sum_{i=1}^{N} \lambda_i = 1,\ \lambda_i \geq 0,\ x_i \in H\right\} = \|\rho^{**}(\mu_f)\| .$$

Proof. To avoid complications with the modular function we derive the first formula from the second using the natural involution in $M(G)$ and the following observation.

Lemma. If $\mu \in M(G)$, then $\|\rho^{**}(\mu^*)\| = \|\lambda^{**}(\mu)\|$.

Proof. Evidently $< \dot{\rho}^{**}(\mu), \psi > = < \mu, \psi \circ \rho >$ for all continuous bounded ψ on $(G/H)_r$. If $\psi \in C_0((G/H)_r)$ then $f'(t) = \overline{\psi(\rho(t^{-1}))}$ is continuous, constant on *left* cosets of H, and $\|f'\|_\infty = \|\psi\|_\infty$; thus there is a continuous f on $(G/H)_\ell$ with $f' = f \circ \lambda$ and $\|f\|_\infty = \|f'\|_\infty$. Hence

$$\begin{aligned} |< \rho^{**}(\mu^*), \psi >| &= |\int \psi(\rho(t))d\mu^*(t)| \\ &= |\int \overline{\psi(\rho(t^{-1}))}d\mu(t)| \\ &= |\int f(\lambda(t))d\mu(t)| = |< \lambda^{**}(\mu), f >| \end{aligned}$$

and so

$$\begin{aligned} \|\rho^{**}(\mu^*)\| &= \sup\{|< \mu^*, \psi \circ \rho >|: \psi \in C_0((G/H)_r), \|\psi\|_\infty = 1\} \\ &\leq \sup\{|< \lambda^{**}(\mu), f >|: \\ &\qquad f \in CB((G/H)_\ell), \|f\|_\infty = 1\} = \|\lambda^{**}(\mu)\| . \end{aligned}$$

Interchanging roles of λ, ρ we can prove $\|\lambda^{**}(\mu^*))\| \leq \|\rho^{**}(\mu)\|$, and (as $(\mu^*)^* = \mu$) conclude $\|\lambda^{**}(\mu)\| \geq \|\rho^{**}(\mu^*)\| \geq \|\lambda^{**}(\mu)\|$.

Q.E.D.

Involution in $M(G)$ is isometric, and $(\lambda\delta_x)^* = \overline{\lambda}\,\delta_{x^{-1}}$, so the lemma shows:

$$\inf\{\|\Sigma\lambda_i \mu_f * \delta_{x_i}\| = \inf\{\|\Sigma \lambda_i \delta_{x_i^{-1}} * \mu_f^*\|\}$$

$$= \|\rho^{**}(\mu_f^*)\| = \|\lambda^{**}(\mu_f)\| ,$$

as desired.

For convex sums ($x_i \in H$) we always have

$$\inf\{\|\Sigma \lambda_i \delta_{x_i} * \mu_f\|\} \geq \|\rho^{**}(\mu_f)\| ;$$

in fact if $\psi \in C_0((G/H)_r)$ with $\|\psi\|_\infty = 1$, then

$$\|\sum_{r=1}^{N} \lambda_i \delta_{x_i} * \mu_f\| \geq |< \sum_{i=1}^{N} \lambda_i \delta_{x_i} * \mu_f, \psi \circ \rho >|$$

$$= |\sum_{i=1}^{N} \lambda_i \int \psi \circ \rho(t) f(x_i^{-1} t) dt|$$

$$= |\sum_{i=1}^{N} \lambda_i \int \psi \circ \rho(x_i t) f(t) dt| .$$

But $\psi \circ \rho$ is constant on the right coset $Ht \supset \{x_i t : i = 1, 2, \ldots, N\}$, so the last sum is just $|\int \psi \circ \rho(t) f(t) dt| = |< \rho^{**}(\mu_f), \psi >|$ and we see

$$\inf\{\|\sum_{i=1}^{N} \lambda_i \delta_{x_i} * \mu_f\|\} \geq \|\rho^{**}(\mu_f)\| .$$

On the other hand consider $L_x(f)(t) = f(xt)$ in $L^1(G)$ and L_x^* the adjoint operator on $L^\infty(G)$. Following [21], if

$$\delta = \inf\{\| \Sigma_i \lambda_i \delta_{x_i} * \mu_f\|\} > 0$$

there is some $f^* \in L^\infty(G)$ with $\|f^*\|_\infty = 1$ such that

$$|< f^*, h >| \geq \text{Re} < f^*, h > \geq \delta$$

for all h in the norm closed convex hull $C_H(f)$ of $\{L_x(f) : x \in H\}$, see [13] V. 2.8; this implies the same inequality:

$$\delta \leq |<\phi, h>| \text{ all } h \in C_H(f)$$

for all ϕ in the weak $*$ closed convex hull $\Sigma(f^*)$ of $\{L_x^*(f^*): x \in H\}$, since it is true for all finite convex sums:

$$|<\sum_{i=1}^{N} \lambda_i L_{x_i}^*(f^*), h>| = |<f^*, \sum_{i=1}^{N} \lambda_i L_{x_i}(h)>| \geq \delta$$

$$\text{all } h \in C_H(f).$$

Thus by definition of δ the set $\Sigma(f^*)$, which obviously lies within the unit ball in L^∞ since $\|f^*\|_\infty = 1$, consists entirely of elements of norm one. Now it is trivial to check that G acts affinely on $\Sigma(f^*)$ and $G \times \Sigma(f^*) \to \Sigma(f^*)$ is continuous. By the fixed point property for G there is some $\sigma^* \in \Sigma(f^*)$ such that $L_x^*(\sigma^*) = \sigma^*$ all $x \in H$. Thus $\|\sigma^*\|_\infty = 1$ and σ^* is constant on *right* cosets of H. Finally we note that $\sigma^* \in \Sigma(f^*)$, so that $h \in C_H(f) \Rightarrow |<\sigma^*, h>| \geq \delta$. Let $\{\psi_i\} \subset CB(G)$ be weak $*$ convergent in L^∞: $\psi_i \to \sigma^*$, with $\|\psi_i\|_\infty = 1$ and ψ_i constant on right cosets of H; let f_i be continuous bounded functions on $(G/H)_r$ with $\psi_i = f_i \circ \rho$. Then $\|f_i\|_\infty = 1$ and $\delta \leq |<\sigma^*, f>| \leftarrow |<\mu_f, f_i \circ \rho>| = |<\rho^{**}(\mu_f), f_i>| \leq \|\rho^{**}(\mu_f)\|$ $\leq \delta$; hence $\|\rho^{**}(\mu_f)\| = \inf\{\|\Sigma \lambda_i \delta_{x_i} * \mu_f\|\}$. Q.E.D.

If $(G/H)_\ell$ admits a measure $d\nu(\xi)$ invariant under the action of G, g: $xH \to gxH$ (such measures exist $\Leftrightarrow$ the modular functions on G, H are related by $\Delta_G|H = \Delta_H$ —see [34]), then we may normalize left Haar measures m_G, m_H on G, H so that

$$\int_{G/H} \left[\int_H f(xt) dm_H(t) \right] d\nu(\xi) = \int f(t) dm_G(t)$$

for all continuous f with compact support. Here we identify $F(x) = \int f(xt) dm_H(t)$, constant on left cosets of H, with a continuous compactly supported function on $(G/H)_\ell$. Then we have a convenient evaluation:

$$\|\lambda^{**}(\mu_f)\| = \int_{G/H} |\int_H f(xt)dm_H(t)|d\nu(\xi)$$

which gives:

$$(*) \quad \inf\{\Sigma_i \lambda_i \mu_f * \delta_{x_i}\} = \int_{G/H} |\int_H f(xt)dm_H(t)|d\nu(\xi) .$$

These formulae are the ones commonly derived in the literature, but the formula in 3.7.2 is valid whether $d\nu(\xi)$ exists or not. We will not prove (*) here; there is a similar formula for $\|\rho^{**}(\mu_f)\|$ when $(G/H)_r$ supports an invariant measure.

In [64] Reiter finally proved that the sort of ergodic property in 3.7.1 is equivalent to property (P_1), hence to amenability, so we have the following theorem.

Theorem 3.7.3. A locally compact group G is amenable $\Longleftrightarrow$ for each $f \in L^1(G)$,

$$\inf\{\| \Sigma_i \lambda_i \delta_{x_i} * \mu_f\|\} = | \int f(t)dt |$$

where the inf is taken over all finite convex sums.

Note: We use left translates here to avoid needless complications in non-unimodular groups; the right-handed result could be proved directly in the same way but it is easier to invoke the natural involutional symmetry of $M(G)$. The study of hulls of left translates is *not* tied to left invariance of the means we use below.

Proof. We have seen (=>) in 3.7.1. Conversely, fix $\phi \in P(G)$ and consider the directed system $J = \{\alpha = (f_1, \ldots, f_N;\ \varepsilon)\}$ with $f_i \in P(G)$, $\varepsilon > 0$, and $N < \infty$, and direct J so $\alpha > \alpha' \iff \{f_i\} \supset \{f_i'\}$ and $0 < \varepsilon < \varepsilon'$. Then for each $\alpha \in J$ we have $< f_i * \phi - \phi, 1 > \ = \ 0$ for $i = 1, 2, \ldots, N$, and by hypothesis there are convex sums of point masses $\{\sigma_1, \ldots, \sigma_N\}$ such that

$$\|(f_1 * \phi - \phi) * \sigma_1\| < \varepsilon$$

$$\|(f_N * \phi - \phi) * \sigma_1 * \ldots * \sigma_N\| < \varepsilon .$$

(Notice that $< f, 1 > \ = 0 \implies < f * \sigma_k, 1 > \ = \ < f, 1 > \cdot < \sigma_k, 1 > \ = 0$ too.) Taking $\sigma_\alpha = \sigma_1 * \cdots * \sigma_N$ we see

$$\|(f_k * \phi - \phi) * \sigma_\alpha\|$$

$$\leq \|(f_k * \phi - \phi) * \sigma_1 * \cdots * \sigma_k\| \cdot \|\sigma_{k+1} * \cdots * \sigma_N\|$$

$$= \|(f_k * \phi - \phi) * \sigma_1 * \cdots * \sigma_k\| < \varepsilon$$

for $k = 1, 2, \ldots, N$. If we now define $\{\phi_\alpha\} \subset P(G)$ with $\phi_\alpha = \phi * \sigma_\alpha$ for each $\alpha \in J$ then $\{\phi_\alpha\}$ evidently converges strongly to topological left invariance and G is amenable. Q.E.D.

Reiter's original result [58] avoids the fixed point property but is complicated; it deals with locally compact abelian G but is actually valid for all G with property $(\mathbf{P}_1)$ if one injects an occasional modular function into the discussion. Glicksberg has given some interesting extensions in [21]; in particular he discusses the action of amenable groups on spaces of functions supported on homogeneous spaces associated with G.

Example. We derive a classical summation formula: if $f \in L^1(G)$, where $G = \mathbf{R}$, then

$$\int_0^1 |\sum_{n=-\infty}^{\infty} f(x+n)|dx = \lim_{N\to\infty} \int_{-\infty}^{\infty} |\frac{1}{2N+1} \sum_{n=-N}^{N} f(x+n)|dx .$$

Consider $H = \mathbf{Z}$ and $d\lambda(\xi)$ normalized left Haar measure on G/H, and let $\Sigma_{fin}(H) = \{\mu \in M(H)\colon \mu = \Sigma\, \lambda_i \delta_{x_i},\ x_i \in H,\ \Sigma\, \lambda_i = 1, \lambda_i \geq 0\}$. Then formula (*), generalizing 3.7.2, gives

$$\int_0^1 |\sum_{n=-\infty}^{\infty} f(x+n)|dx = \inf\{\|\mu * \mu_f\|\colon \mu \in \Sigma_{fin}(H)\} = \delta .$$

On the other hand we show

$$\delta = \lim_{N\to\infty} \int_{-\infty}^{\infty} |\frac{1}{2N+1} \sum_{n=-N}^{N} f(x+n)|dx .$$

In fact if $\mu_N = \frac{1}{2N+1} \Sigma_{n=-N}^{N} \delta_n$, the right side is just $\lim_{N\to\infty} \|\mu_N * \mu_f\|$; obviously $\delta = \inf\{\|\mu * \mu_f\|\} \leq \|\mu_N * \mu_f\|$ for $N = 1, 2, \ldots$. If $\mu \in \Sigma_{fin}(H)$ is chosen so $\delta \leq \|\mu * \mu_f\| < \delta + \varepsilon$, then $\|\mu_N * \mu * \mu_f\| < \delta + \varepsilon$, while

$$\|\mu_N * \mu * \mu_f - \mu_N * \mu_f\| \leq \|\mu_N * \mu - \mu_N\| \cdot \|\mu_f\| \to 0$$

since $\{\mu_N\}$ is clearly strongly convergent to left invariance in H. We conclude that $\lim_{N\to\infty} \|\mu_N * \mu_f\| = \delta$ as required.

Formula (*) and similar considerations about nets strongly convergent to left invariance on the subgroup H give numerous formulae of this sort. If $G = H$ we get:

$$\lim_{T\to\infty} \int_{-\infty}^{\infty} |\frac{1}{2T} \int_{-T}^{T} f(x+t)dt|dx = |\int_{-\infty}^{\infty} f(x)dx| .$$

§3.8. WEAKLY ALMOST PERIODIC SEMIGROUPS OF OPERATORS

This family of results does not involve amenability exactly as we have defined it, but rather demands existence of invariant means on the "weakly almost periodic" functions $W(S)$ on a *topological semigroup* S, by which we mean a semi-group with *separately continuous* multiplication. We will not insist that S be a group or locally compact, so the existence of these means is not a consequence of the considerations in 3.1.

Let X be a Banach space $\mathcal{B}(X)$ the continuous linear operators on X, and in $\mathcal{B}(X)$ define the *(wo)-topology* so that

$$A_j \xrightarrow{(wo)} A \iff \langle A_j x - A,\ x^* \rangle \to 0$$

$$\text{for all } x \in X,\ x^* \in X^*.$$

Let $\{T_s\colon s \in S \text{ (an index set)}\}$ be a semigroup of operators in $\mathcal{B}(X)$. It is separately continuous in the (wo)-topology; such a semigroup is said to be *weakly almost periodic* (WAP) if the set $\{T_s\}$ is relatively compact (closure is compact) in $(\mathcal{B}(X), (wo))$.[6, 7] It is not difficult to verify the following equivalent condition (proof $(\Longrightarrow)$ is trivial; the details of $(\Longleftarrow)$ are straightforward—see [23], Theorem 3.1 for a self-contained discussion).

---

(6) The following simple lemma facilitates discussion of sets whose closure is compact. It may fail to be true in general (non-uniform) topological spaces. *Lemma.* In any topological vector space X, a set $E \subset X$ has compact closure $E^- \iff$ every net $\{e_j\} \subset E$ has a limit point in X. The proof is unexceptional.

(7) *Almost periodicity* is also interesting: this means $\{T_s\}$ is relatively compact in $(\mathcal{B}(X), (so))$ where (so) topology is defined so that

$$A_j \xrightarrow{(so)} A \iff \|A_j x - Ax\| \to 0 \quad \text{for all } x \in X.$$

Lemma 3.8.1. A semigroup of operators $\{T_s\} \subset \mathcal{B}(X)$ on a Banach space X is WAP $\Longleftrightarrow$ for any $x \in X$ the weak closure of the orbit $O(x) = \{T_s(x): s \in S\}$ is a weakly compact set in the Banach space X.

If $\{T_s: s \in S\}$ is WAP semigroup of operators on X, then $\sup\{\|T_s\|: s \in S\} < \infty$. In fact, for $x \in X$ the weak closure $O(x)^- \subset X$ is weakly compact while every function $y \to \langle y, x^* \rangle$ $(x^* \in X^*)$ is a continuous function with respect to the weak topology in X; thus $\sup\{\langle T_s(x), x^* \rangle: s \in S\} < \infty$ for each $x \in X$, $x^* \in X^*$ and the uniform boundedness principle applies ([13], II. 3.21).

The *weakly almost periodic functions* $W(S)$ on a topological semigroup S are those $f \in CB(S)$ whose orbit $O(f) = \{R_s f: s \in S\}$ under right translation operators $R_s f(t) = f(ts)$ is relatively compact in $(CB(S), (wk))$ where (wk) indicates the weak topology in $CB(S)$. It is trivial to show that $W(S)$ is a norm closed, two-sided invariant linear subspace in $CB(S)$. The class of functions just defined should really be called right weakly almost periodic; there is a similar notion using left translates $L_s f(t) = f(st)$. If G is a topological group, it is well known that the left, right, and two-sided notions coincide; more surprisingly (in view of the asymmetry of semigroups) left and right weak almost periodicity are equivalent for any separately continuous topological semigroup, as was shown by Grothendieck [30]. We shall not reproduce this proof here.

If S is a topological semigroup the semigroup of right translation operators $\{R_s\}$ on $W(S)$ is the classic example of a WAP semi-group of operators: if $f \in W(S)$ its orbit

$$O(f) = \{R_s f:\ s \in S\}$$

has closures with respect to the weak topologies in $CB(S)$ a
$W(S)$. The definition of $f \in W(S)$ insures that the closure of
$O(f)$ with respect to the weak topology in $CB(S)$ is compact
this topology. An elementary application of the Hahn-Banac
extension theorem shows that these weak topologies coincid
on $W(S)$: in fact, $W(S)$ is a norm closed convex set in $CB(S)$
so it is closed in the weak topology of $CB(S)$—see [13] V.3.
Thus each orbit has weakly compact weak closure in $W(S)$, s
Lemma 3.8.1 applies to show $\{R_s\}$ is a WAP semigroup of op
erators on $X = W(S)$.

If S is an abstract WAP semigroup of operators on Bana
space X there is still a connection with weakly almost peric
ic functions: take T the identity map in the following situation.

Lemma 3.8.2. Let S be a topological semigroup and $T: S$ -
$(\mathcal{B}(X), (wo))$ a continuous map representing S $(T_{st} = T_s \circ T_t)$
as a WAP semigroup of operators $\{T_s\}$ on X. Then all representation functions, $f(s) = < T_s(x), x^* >$ for $x \in X$, $x^* \in X^*$
are weakly almost periodic on S.

Proof. We have seen that $\{T_s\}$ is uniformly bounded in opera
tor norm, so the representation functions are all in $CB(S)$, cc
tinuity reflecting the continuity of the map T. Fix $x^* \in X^*$
and define $\Phi: x \to CB(S)$ so $\Phi x(s) = < T_s x, x^* >$; Φ is line
norm continuous, and its adjoint $\Phi^*: CB(S)^* \to X^*$ has
$< \Phi^* m, x > = < m, \Phi x >$. Consider any net $\{R_{s_j}:\ j \in J\}$ of
right translations, then Φx is a typical representation functi

and $\{R_{s_j}(\Phi x)\}$ a typical net in $O(\Phi x) \subset CB(S)$. We show that such a net has a weak limit point in $CB(S)$, which proves that $\Phi x \in W(S)$.

We have

$$R_{s_j}(\Phi x)(s) = \Phi x(ss_j) = \langle T_s T_{s_j}(x), x^* \rangle = \Phi(T_{s_j}(x))(s),$$

which $\Longrightarrow R_{s_j}(\Phi x) = \Phi(T_{s_j}(x))$ and

$$\langle R_{s_j}(\Phi x), m \rangle = \langle \Phi(T_{s_j}(x)), m \rangle = \langle T_{s_j}(x), \Phi^* m \rangle$$

for all $m \in CB^*$. We know $O(x)$ is relatively compact in $(X, (wk))$ so we may select a subnet, if necessary, and assume

$$T_{s_j}(x) \xrightarrow{(wk)} x_\infty \in X \text{ in } X.$$

Then for all $m \in CB^*$:

$$\langle T_{s_j}(x), \Phi^* m \rangle \to \langle x_\infty, \Phi^* m \rangle = \langle \Phi(x_\infty), m \rangle,$$

which $\Longrightarrow R_{s_j}(\Phi x) \to \Phi(x_\infty)$ in $(CB(S), (wk))$. Q.E.D.

Weak relative compactness[(8)] of $O(x) = \{T_s(x): s \in S\} \subset X$ in the above context allows us to construct weak vector valued integrals $\int T_s x \, dm(s)$ with respect to any mean m on $W(S)$, as follows.

---

(8) If the weak closure of $O(x)$ is weakly compact, then so is the weakly closed convex hull $C(x)$ of $O(x)$ [this is the same as the norm closed convex hull, by Hahn-Banach], as we have indicated in footnote (3). It is the weak compactness of $C(x)$ which we need below.

Lemma 3.8.3, Let $T\colon\ S \to (\mathcal{B}(X), (wo))$ be a continuous repr sentation of topological semi-group S as a WAP semigroup o operators,and let m be a mean on $W(S)$. Corresponding to each $x \in X$ there is a unique element of X, denoted $I(x) = \int T_s x\, dm(s)$, such that

$$(1) \qquad < I(x), x^* > \;=\; \int < T_s x, x^* > dm(s) \quad \text{all } x^* \in X^*$$

(the right hand integral being interpreted via 3.8.2). Further-more if $C(x)$ is the weakly closed convex hull of $O(x) = \{T_s(x)\colon\ s \in S\}$ we have $I\colon\ X \to X$ a bounded linear mapping with $I(x) \in C(x)$, and if m is a left [right] invariant mean on $W(G)$ we have

$$(2) \qquad T_s(Ix) = Ix \quad [I(T_s x) = Ix]$$

for all $s \in S,\ x \in X$.

Proof. Formula (1) determines a unique element of X^{**}. Th point evaluation functionals $\{\delta_s\colon\ s \in S\}$ in $W(S)^*$ have conve hull Σ_{fin} weak$*$ dense in the set Σ of all means in $W(S)^*$: otherwise, there would be some $m' \in W(S)^*$ with $\|m'\| = 1$ and (see [13], V. 2.10) some $f \in W(S)$ such that $\mathrm{Re} < m', f > \;\geq \varepsilon + \mathrm{Re} < \delta_s, f > \;= \mathrm{Re}\, f(s)$, for all $s \in S$ and some $\varepsilon > 0$, so that $< m', \mathrm{Re}\, f > \;\geq \varepsilon + \sup\{\mathrm{Re}\, f\}$, a contradiction. Thus, let $\{\sigma_j\} \subset \Sigma_{fin}$ be chosen so $\sigma_j \to m$ weak$*$; then for fixed x, x^*:

$$\int < T_s x, x^* > dm(s) \;\leftarrow\; \int < T_s x, x^* > d\sigma_j(s)$$

$$= \Sigma_{s \in S}\, \lambda(j, s) < T_s x, x^* > \;=\; < \Sigma\, \lambda(j, s) T_s x, x^* > ,$$

where $\sigma_j = \Sigma_{s \in S} \lambda(j, s)\delta_s$ with $\lambda(j, s) \geq 0$ and $\Sigma_{s \in S}\lambda(j, s) = 1$ a finite sum. But $x_j = \Sigma \lambda(j, s)T_s x \in C(x)$, which is weakly compact, so we may take a subnet to get weak convergence $x_{j(k)} \to x_\infty \in C(x)$, while we still have $\sigma_{j(k)} \to m$ weak $*$ in $W(G)^*$. Hence

$$< I(x), x^* > \; = \; < x_\infty, x^* > \quad \text{for all } x^* \in X^*,$$

so $I(x) = x_\infty \in X$. The other properties of $I: X \to X$ are immediate from formula (1) and the uniform boundedness of $\{T_s : s \in S\}$. Q.E.D.

Note: To make this construction work we only need a mean m on the norm closed subspace of $W(S)$ spanned by all representation functions and all constants. Dixmier [11] gives a result of this sort for discrete *amenable* semigroups (invariant means on all of $B(S)$). We have indicated in Section 3.1 that *every* locally compact group G has a (unique, two-sided) invariant mean on $W(G)$ so this construction, and the following theorem, applies to any (wo)-continuous representation of G in a reflexive Banach space.

Theorem 3.8.4. Let S be a topological semigroup which admits left and right invariant means m_ℓ and m_r on $W(S)$. Let $T: S \to (\mathcal{B}(X), (wo))$ be a continuous representation of S as a WAP semigroup of operators in Banach space X. If $X_0 = \{x \in X: T_s(x) = x, \text{ all } s \in S\}$ and X_1 is the closed subspace spanned by $\{T_s(x) - x: s \in S, x \in X\}$ then

(1) $X_0 \cap X_1 = (0)$, $X = X_0 \oplus X_1$, and the projection $E: X \to X_0$ is a continuous idempotent linear operator.

(2) For each $x \in X$ the norm closed convex hull $C(x)$ of $O(x) = \{T_s(x)\}$ meets X_0 in a unique point which is precisely $E(x)$.

Proof. We have well defined weak vector valued integrals as in 3.8.3:

$$R(x) = \int T_s(x)dm_r(s) \qquad L(x) = \int T_s(x)dm_\ell(s)$$

with $R(x) \in C(x)$ and $R(T_s x) = R(x)$ [resp. $T_s(Lx) = L(x)$]. First note that $L(X) \subset X_0$ since $T_s(Lx) = L(x)$; but if $x \in X_0$ then $< Lx, x^* > = \int < T_s x, x^* > dm_\ell(x) = < x, x^* >$ for all $x^* \in X^*$, which $\Longrightarrow Lx = x$ and $X_0 \subset L(X)$, proving $L(X) = X_0$. Furthermore,

$$\begin{aligned} < L^2 x, x^* > &= \int < T_s(Lx), x^* > dm_\ell(s) \\ &= \int < Lx, x^* > dm_\ell(s) = < Lx, x^* > \end{aligned}$$

so $L^2 = L$ and L is the projection of X onto X_0.

Actually, $R(x) = L(x)$ for all $x \in X$: in fact

$$R(\sum_{i=1}^{N} \lambda_i T_{s_i} x) = \sum_{i=1}^{N} \lambda_i R(T_{s_i} x) = R(x)$$

for all convex sums, so $R(z) = R(x)$ for all $z \in C(x)$. But $L(x) \in C(x)$, so $R(x) = R(L(x)) = \int T_s(Lx)dm_r(s)$; the latter is just $L(x)$ since $L(x) \in X_0 \Longrightarrow T_s(Lx) = Lx$ for all $s \in S$. Thus we see $R = L$ and $T_s L(x) = LT_s(x) = L(x)$ all $s \in S$. Now consider $\mathrm{Ker}\,(L)$: evidently $X = X_0 \oplus \mathrm{Ker}\,(L)$. Now if $x \in \mathrm{Ker}\,(L)$ then $LT_s(x) = Lx = 0$, so $T_s(x) \in \mathrm{Ker}\,(L)$ for all $s \in S$ and $C(x) \subset \mathrm{Ker}\,(L)$. For arbitrary $x \in X$ we have

$x - L(x) \in \text{Ker}(L)$, so $C(x-Lx) \subset \text{Ker}(L)$; but

$$C(x) = C((x-Lx)+Lx) = Lx + C(x-Lx) \subset Lx + \text{Ker}(L).$$

But $Lx \in C(x) \cap X_0$ and if there were two points $p, g \in C(x) \cap X_0$ we may write $p = Lx + p'$, $q = Lx + q'$ with $p', q' \in \text{Ker}(L)$, so $p-q = p'-q'$; this difference lies within $X_0 \cap \text{Ker}(L) = (0)$, so we see $C(x) \cap X_0 = \{L(x)\}$ as required.

Finally, for any $x \in X$ we have $L(x - T_s x) = Lx - L(T_s x) = 0$ so $X_1 \subset \text{Ker}(L)$. Conversely if $z \in \text{Ker}(L)$, then $0 = L(z) \in C(z)$ and there are convex sums with

$$\left\| \sum_{i=1}^{N} \lambda_i T_{s_i}(z) \right\| < \varepsilon \ ;$$

thus

$$z = \left(z - \sum_{i=1}^{N} \lambda_i T_{s_i}(z) \right) + \sum_{i=1}^{N} \lambda_i T_{s_i}(z)$$

$$= \sum_{i=1}^{N} \lambda_i (z - T_{s_i}(z)) + z_1$$

where $\|z_1\| < \varepsilon$. Thus z is adherent to the closed linear span of $\{T_s(x) - x\}$ so $\text{Ker}(L) \subset X_1$, and we see $X_1 = \text{Ker}(L)$.

Q.E.D.

Again it is only necessary to have m_ℓ, m_r left and right invariant means on the closed invariant subspace in $W(S)$ generated by the constant functions and the representations functions $s \to \langle T_s x, x^* \rangle$. Below we give another situation where existence of such means for a WAP semigroup of operators on X gives a direct sum decomposition of X; this is typical of the results in [23].

Theorem 3.8.5. Let S be a topological semigroup which admits left and right invariant means m_ℓ and m_r on $W(S)$, let $T: S \to (\mathcal{B}(X), (wo))$ be a continuous representation of S as a WAP semigroup of operators on Banach space X, and let $\overline{S}$ be the (wo) closure of $\{T_s : s \in S\}$—a compact semigroup of operators in $(\mathcal{B}(X), (wo))$. Define the subsets of X:

(i) $X_r = \{x \in X$: for each $U \in \overline{S}$, there exists some $V \in \overline{S}$ such that $VUx = x\}$;

(ii) $X_f = \{x \in X$: 0 is weakly adherent to the orbit $O(x)\}$,

which we refer to as the "reversible vectors" and the "flight vectors" respectively. Then X_r, X_f are closed $\overline{S}$-invariant subspaces of X with $X = X_r \oplus X_f$.

We will not prove this; the reader is directed to [23] for full details. The following example, drawn from [23], will help understand the meaning of 3.8.5.

Example. Let S be a compact (separately continuous) topological semigroup. It is not hard to show that $W(S) = CB(S)$ in this situation. Then consider the representation $R_s f(t) = f(ts)$ on $X = CB(S)$; we have seen this is a WAP semigroup of operators on X. From the elementary structure theory of compact semigroups we see that $CB(S)$ has a unique 2-sided invariant mean m if it has left, right invariant means m_ℓ, m_r. Clearly 3.8.5 applies, giving $W(S) = W(S)_f \oplus W(S)_r$, but it is very interesting to interpret $W(S)_f$, $W(S)_r$ in terms of the unique invariant mean m. It can be shown that:

$$W(S)_f = \{x \in W(S): \langle m, |x|^2 \rangle = 0\}$$

$$W(S)_r = AP(S), \text{ the almost periodic functions on } S,$$

so $W(S) = AP(S) \oplus \{x \in W(S) : \langle m, |x|^2 \rangle = 0\}$. When S is a locally compact abelian group, this is a well known result due to Eberlein.

For further discussion of WAP semigroups see [22], [23], [39], [40]. Interesting applications to the theory of group representations may be found in [22].

APPENDIX 1

NONUNIQUENESS OF INVARIANT MEANS

E. Granirer has studied the uniqueness question for left invariant means on semigroups in [26], extending the work of Day in [8]. His work includes the following definitive nonuniqueness theorem for discrete groups.

Theorem A.1.1. If G is any infinite amenable group, there exist many left invariant means on $B(G)$.

Recently R. Kaufmann [42] introduced some probabilistic notions to give a new proof of this result for abelian groups. We adapt these ideas to give a proof of A.1.1. The result is proved in a series of lemmas (regarding the first lemma below, cf. [52], pp. 256-257).

Lemma A.1.2. Let G be amenable and let $S, T \subset G$. In order that there be a LIM m such that $m(\chi_S) = 1$, $m(\chi_T) = 0$ it is necessary and sufficient that for every finite set $F \subset G$ there exist $x \in G$ such that $Fx \subset S$ and $Fx \cap T \neq \emptyset$.

Proof. For necessity, let $F = \{a_1, \ldots, a_N\}$ and write $m(E) = m(\chi_E)$. Since $m(S) = m(G) = 1$ it is easily seen that

$$m\left(\bigcap_{i=1}^{N} a_i^{-1}S\right) = 1 ,$$

so

$$\left(\bigcap_{i=1}^{N} a_i^{-1}S\right)\setminus\left(\bigcup_{i=1}^{N} a_i^{-1}T\right)$$

has measure 1 and is thus nonempty; if x lies in this set, clearly $Fx \subset S$ and $Fx \cap T = \emptyset$.

Conversely since G is amenable the Følner condition (see section 3.6) is valid, hence there is a net $\{F_j : j \in J\}$ of finite nonempty subsets with the property

$$\frac{|xF_j \,\Delta\, F_j|}{|F_j|} \to 0 \quad \text{all } x \in G,$$

where $|E|$ is the cardinality of $E \subset G$. Select $g_j \in G$ so that $F_jg_j \subset S$, $F_jg_j \cap T = \emptyset$ all $j \in J$. Then

$$\frac{|x(F_jg_j) \,\Delta\, (F_jg_j)|}{|F_jg_j|} = \frac{|xF_j \,\Delta\, F_j|}{|F_j|} \to 0 \quad \text{all } x \in G,$$

and if ϕ_j is the normalized characteristic function of F_jg_j ($\phi_j \in \ell^1(G)$) this means that $\{\phi_j\}$ is strongly convergent to left invariance: $\|{}_x\phi_j - \phi_j\|_1 \to 0$, all $x \in G$. Thus if we regard $\{\phi_j\} \subset B(G)^*$, every weak $*$ limit point m is a LIM on $B(G)$. But (taking a subnet weak $*$ convergent to m)

$$< m, \chi_T > \leftarrow < \phi_j, \chi_T > = \frac{1}{|F_j|} \Sigma\{\chi_{F_jg_j}(p)\chi_T(p) : p \in G\} = 0$$

$$< m, \chi_S > \leftarrow < \phi_j, \chi_S > = \frac{1}{|F_j|} \Sigma\{\chi_{F_jg_j}(p)\chi_S(p) : p \in G\} = 1$$

since $F_jg_j \subset S$. Q.E.D.

Now consider the space 2^G of all subsets of G, identified in the obvious way with the set of all functions $f\colon G \to \{0, 1\}$ with the product topology: 2^G is metrizable if G is countable. Give the two point space $\{0, 1\}$ the usual probability measure and let μ be the product probability measure on 2^G (well defined even if G is uncountable); this measure is defined at least on the Baire sets (generated by compact G_δ sets) and will be regarded as a Baire measure. We note that if $F \subset G$, $|F| < \infty$ then $\mu\{S \in 2^G\colon F \subset S\} = 2^{-|F|}$.

Lemma A.1.3. Let G be any group, $F \subset G$ a finite subset of G, and $U = \{(S, T) \in 2^G \times 2^G\colon (\exists x)(Fx \subset S \text{ and } Fx \cap T = \emptyset)\}$. Then U is an open set; if G is infinite, then U is dense in $2^G \times 2^G$ and contains an open Baire set of $\mu \times \mu$ measure 1.

Proof: U is exhibited as a union of open sets. If $|G| = \infty$ and $\tilde{F} = FF^{-1} \cup F^{-1}F$ then we may choose a sequence $\{x_k\colon k = 1, 2 \ldots\}$ such that $x_i^{-1}x_j \notin \tilde{F}$ and $x_j^{-1}x_i \notin \tilde{F}$ if $i \neq j$. Obviously the complement of U lies within the closed Baire set

$$\bigcap_{i=1}^{\infty} \{(S, T) \in 2^G \times 2^G\colon Fx_i \not\subset S \text{ or } Fx_i \cap T \neq \emptyset\} = \bigcap_{i=1}^{\infty} E_i .$$

If we show this set has $\mu \times \mu$ measure zero, it follows that its complement (hence also U) is dense since any open nonempty set in $2^G \times 2^G$ contains an open Baire set with positive measure, and our theorem is proved. The general idea in proving $\mu \times \mu(\bigcup_{i=1}^{\infty} E_i) = 0$ is that the sets E_i can be shown to be "jointly independent" ([31], section 45): i.e.,

$$\mu \times \mu\left(\bigcap_{i=1}^{\infty} E_i\right) = \prod_{i=1}^{\infty} \mu \times \mu(E_i) ;$$

but it is clear that $\mu \times \mu(E_i) = (1 - 2^{-2|F|})$ for $i = 1, 2, \dots$, and we are done. Here is a detailed proof that $\bigcup_{i=1}^{\infty} E_i$ is a null set, with probabilistic considerations suppressed.

Index $G = \{x_a \colon a \in I\}$, let $P = \{0, 1\}$ with ν the usual probability measure on P, and let $f_a(S, T) = \chi_S(x_a)$, $g_a(S, T) = \chi_T(x_a)$. Then the map $\Phi = \Pi f_a \times \Pi g_a \colon 2^G \times 2^G \to \Pi P_a \times \Pi P_a$ preserves measurability and carries $\mu \times \mu$ to the product measure $\bar{\nu} \times \bar{\nu} = \Pi \nu_a \times \Pi \nu_a$. Writing elements of $\Pi P_a \times \Pi P_a$ as $(\xi; \eta) = (\xi(a); \eta(a))$ we see that

$$E_i = (\{S \colon Fx_i \not\subset S\} \times 2^G) \cup (2^G \times \{T \colon T \cap Fx_j \neq \emptyset\})$$

so that

$$\Phi(E_i) = (\{\xi \colon \xi(a) \not\equiv 1 \text{ on } Fx_i\} \times \Pi P_a)$$
$$\cup\ (\Pi P_a \times \{\eta \colon \eta(a) \not\equiv 0 \text{ on } Fx_i\})$$

in $\Pi P_a \times \Pi P_a$. Clearly we may write

$$\Phi(E_1 \cap \dots \cap E_N) = A_N \cup B_N$$
$$= (\{\xi \colon \text{for } 1 \leq k \leq N,\ \xi(a) \not\equiv 1 \text{ on } Fx_i\} \times \Pi P_a)$$
$$\cup\ (\Pi P_a \times \{\eta \colon \text{for } 1 \leq k \leq N,\ \eta(a) \not\equiv 0 \text{ on } Fx_k\})$$

so that

$$\mu \times \mu\Big(\bigcap_{i=1}^{N} E_i\Big) = \bar{\nu} \times \bar{\nu}\Big(\Phi\Big(\bigcap_{i=1}^{N} E_i\Big)\Big) \leq \bar{\nu} \times \bar{\nu}(A_N) + \bar{\nu} \times \bar{\nu}(B_N)\,.$$

But $\bar{\nu} \times \bar{\nu}(A_N) = \bar{\nu}(\{\xi \in \Pi P_a \colon \text{for } 1 \leq k \leq N,\ \xi(a) \not\equiv 1 \text{ on } Fx_k\})$, and the latter is readily seen to be

$$(2^{-|F|}(2^{|F|} - 1))^N = (1 - 2^{-|F|})^N,$$

so that

$$\mu \times \mu\left(\bigcap_{i=1}^{N} E_i \right) \leq 2(1-2^{-|F|})^N \to 0$$

as required. Q.E.D.

Corollary A.1.4. If G is a countably infinite amenable group then for $\mu \times \mu$ almost every (S, T) there are left invariant means m, m' with $m(S) = 1$, $m(T) = 0$ and $m'(S) = 0$, $m'(T) = 1$. The pairs (S, T) with this property form a dense G_δ-set in $2^G \times 2^G$.

Proof. The finite subsets of G are denumerable, so the desired set E is the intersection of the G_δ-sets:

$$\bigcap_{|F|<\infty} \left\{ \bigcup_{x \in G} \{(S, T) : Fx \subset S \text{ and } Fx \cap T = \emptyset\} \right\}$$

$$\bigcap_{|F|<\infty} \left\{ \bigcup_{x \in G} \{(S, T) : Fx \cap S = \emptyset \text{ and } Fx \subset T\} \right\}$$

Each has $\mu \times \mu$ measure 1 and so does their intersection E, thus E is dense. Now apply A.1.2. Q.E.D.

If $|G| = \infty$ (not necessarily countable) let $\{a_n : n = 1, 2, \ldots\}$ be distinct elements and $H \subset G$ the denumerable group generated by $\{a_n\}$; apply A.1.4. and the following result to prove the main theorem in full generality.

Lemma A.1.5. If G is amenable and H any subgroup (necessarily amenable) then G has distinct left invariant means if H does.

Proof. Let m be a fixed LIM on $B(G)$ and m_1, m_2 left invariant means on $B(H)$. For $f \in B(G)$ define $A_i f \in B(G)$:

$$A_i f(t) = \int_H f(th) dm_i(h)$$

(constant on left cosets tH). Then define

$$M_i(f) = < m, A_i f > .$$

Evidently M_i is a LIM on $B(G)$ for $i = 1, 2$ since we have $_x(A_i f) = A_i(_x f)$ for $x \in G$. Now let T be a transversal for the right cosets of H and let $x = \tau(x)\eta(x) \in T \cdot H$ be the unique factorization of $x \in G$. Extend each $g \in B(H)$ to $\hat{g} \in B(G)$: $\hat{g}(x) = g(\eta(x))$ all $x \in G$. One readily verifies that $M_i(\hat{g}) = m_i(g)$, thus $m_1 \neq m_2$ on $B(H)$ implies $M_1 \neq M_2$ on $B(G)$. Q.E.D.

APPENDIX 2

THE RYLL-NARDZEWSKI FIXED POINT THEOREM

The following discussion is adapted from Namioka-Asplund [56] and some unpublished remarks of **J. L. Kelley** and M. Rieffel.

Theorem A.2.1. Let E be a separable Banach space, let K be a weakly compact convex set in E, and let $\varepsilon > 0$. Then there is a closed convex subset $C \subset K$ such that $\operatorname{diam}(K \setminus C) < \varepsilon$.

Remarks: Weak and norm closures are the same for convex sets. Our theorem shows that such sets as K are *dentable*: for any $\varepsilon > 0$ there is a subset $L \subset K$ with $\operatorname{diam}(L) < \varepsilon$, $K \setminus L$ a convex subset $(K \setminus L)^{-} \neq K$ (norm and weak closures are the same).

Proof. Write $N_{\varepsilon} = \{x \colon \|x\| \leq \frac{\varepsilon}{4}\}$ and let P be the weak closure of the set of extreme points in K. Cover P with countably many sets $N_{\varepsilon} + x_i$, $x_i \in P$; as P is weakly compact, hence second category in itself, there is a point $x \in P$ and a weakly open neighborhood W of x such that

$$P \cap (N_{\varepsilon} + x) \supset W \cap P \neq \emptyset \ .$$

Let K_1, K_2 be the weakly closed convex hull $(K_i \subset K)$ of

$(P \setminus W)$, $W \cap P$ respectively. Clearly K is the convex hull of weakly compact sets K_1, K_2 by Krein-Milman theorem; furthe more $K_1 \neq K$ because the extreme points of K_1 lie within $P \setminus W$ (see [44], 15.2). Now define

$$C_r = \{tk_1 + (1-t)k_2 \colon k_i \in K_i, r \leq t \leq 1\} \text{ for } 0 \leq r \leq 1.$$

Clearly the C_r are weakly compact and increase as $r \to 0_+$ wi $C_0 = K$, $C_1 = K_1$; it is not too hard to see that C_r is also co vex. Finally we note that $C_r \neq K$ for all $0 < r \leq 1$, for if C_r K, then each extreme point z of K has the form

$$z = \lambda x_1 + (1-\lambda)x_2 \text{ for } x_i \in K_i, \ \lambda \in [r, 1]$$

hence $z = x_1 \in K_1$, contradicting $K_1 \neq K$. Notice that if $y \in K \setminus C_r$ then y has form $y = \lambda x_1 + (1-\lambda)x_2$ where $x_i \in K_i$ $\lambda \in [0, r]$ and

$$\|y - x_2\| = \|\lambda x_1 + (1-\lambda)x_2 - x_2\| = |\lambda| \, \|x_1 - x_2\|$$

$$\leq r\|x_1 - x_2\|,$$

so each $y \in K \setminus C_r$ lies within distance $r \cdot \operatorname{diam}(K_2)$ of K_2. But $\operatorname{diam}(K_2) \leq \frac{\varepsilon}{2}$, thus as $r \to 0_+$, C_r has the desired property $\operatorname{diam}(K \setminus C_r) < \varepsilon$. Q.E.D

Theorem A.2.2. Let E be a Banach space, G a (discrete) semigroup acting affinely on some weakly compact convex se $K \subset E$. Assume G is distal relative to the norm topology in E. Then there is a fixed point for G in K.

Proof. G may be assumed countable, for if $J = \{S_a\}$ is the family of all countable subsemigroups in G, and F_a the set of all points in K left fixed by the elements of S_a, then F_a is weakly closed (hence compact) and $\bigcap_{i=1}^{N} F_{a_i}$ includes all points left fixed by the (countable) semigroup generated by $a_1, \ldots, a_N$. Thus if $F_a \neq \emptyset$ all S_a, we have $\bigcap_{a \in J} F_a \neq \emptyset$ and this set consists of fixed points for G. We may also assume E is separable: otherwise let $x \in K$, replace E with the closed subspace E' generated by $\{Tx: T \in G\}$ and K by the weakly closed convex hull $K' \subset E' \subset E$ of $\{Tx: T \in G\}$. The weak topology in E' coincides with the weak topology inherited from E so K' is weakly compact in E'.

We may assume that K has no nonempty proper closed convex subsets stable under the action of G, because the intersection of the members in a chain of such sets is a set of the same type. But such minimality forces K to be a single point. In fact if there are points $x, y \in K$ with $x \neq y$, we have $\inf\{\|Tx - Ty\|: T \in G\} = \delta > 0$. As K is dentable we may find a subset $L \subset K$ such that $K \setminus L$ is convex, $(K \setminus L)^- \neq K$, and $\operatorname{diam}(L) < \frac{\delta}{2}$. Notice that $T(\frac{1}{2}(x+y)) \notin L$, for otherwise both $T(x)$, $T(y)$ would belong to the convex set $K \setminus L$, which is absurd. Thus the closed convex hull of

$$\{T(\tfrac{x+y}{2}): T \in G\} \subset (K \setminus L)^-$$

is stable under G, violating minimality of K. Q.E.D.

APPENDIX 3

THE EQUIVALENCE OF VARIOUS TYPES OF INVARIANT MEANS

(Another Viewpoint)

Here is an alternative proof of the key lemma used in proving that the various definitions of amenability coincide for locally compact groups. The proof uses weak vector-valued integrals to verify ones intuitive feeling that left invariance of a mean must imply topological left invariance.

Lemma 2.2.2. If m is a LIM on $UCB(G)$ for a locally compact group G, then m is also a topological LIM for $UCB(G)$.

Proof. It obviously suffices to show that $m(\phi * f) = m(f)$ for all $\phi \in P(G)$ with compact support. Let $\phi \in P(G)$, $f \in UCB(G)$. The vector valued mapping $F\colon x \to {}_x f$ is continuous from G into $UCB(G)$; in any Banach space, such as $UCB(G)$, the norm closed convex hull of any norm compact set (e.g. the F-image of supp(ϕ)) is norm compact (see [3]). As is well known there must be an element in $UCB(G)$, the weak vector-valued integral $\int_x f \, d\mu_\phi(x)$ with respect to the measure $d\mu_\phi(x) = \phi(x)dx$, defined by the relation

$$< m, \int_x f \, d\mu_\phi(x) > \; = \int < m, {}_x f > d\mu_\phi(x)$$

$$= \int < m, {}_x f > \phi(x) dx$$

for all $m \in UCB^*$; see [4], pp. 79-89. If m is a LIM on $UCB(G)$ then clearly we have

$$(1)\quad < m, \int {}_xf\, d\mu_\phi(x) > \;=\; < m, f > \cdot \int \phi(x)dx \;=\; < m, f > .$$

Now the point evaluation functionals m_t: $f \to f(t)$ for $t \in G$ obviously have weak $*$ dense linear span in $UCB(G)^*$, and for each of these functionals we have

$$\begin{aligned} < m_t, \int {}_xf\, d\mu_\phi(x) > &= \int \phi(x) < m_t, {}_xf > dx \\ &= \int \phi(x) f(x^{-1}t)dx \\ &= \phi * f(t) \;=\; < m_t, \phi * f > , \end{aligned}$$

so that $\phi * f = \int {}_xf\, d\mu_\phi(x)$ and (1) gives

$$< m, \phi * f > \;=\; < m, f > .$$

Q.E.D.

REFERENCES

[1] S. Banach, "Sur le probleme de mesure," *Fund. Math.* *4* (1923), pp. 7-33.

[2] S. Banach and A. Tarski, "Sur la decomposition des ensembles de points en partes respectivement congruentes," *Fund. Math. 6* (1924), pp. 244-277.

[3] N. Bourbaki, *Espaces vectoriels topologiques, Ch. I-II,* Hermann Cie (1189), 1953, Paris.

[4] ———, *Integration, Ch. I-IV,* Hermann Cie (1175), 1952, Paris.

[5] R. Burckel, *The Ryll-Nardzewski fixed point theorem,* (to appear).

[6] M. Day, "Fixed point theorems for compact convex sets," *Ill. J. Math. 5* (1961), pp. 585-589.

[7] ———, Correction to my paper "Fixed point theorems for compact convex sets," *Ill. J. Math. 8* (1964), p. 713.

[8] ———, "Amenable semigroups," *Ill. J. Math. 1* (1957). pp. 509-544.

[9] ———, "Convolutions, means, and spectra," *Ill. J. Math.* *8* (1964), pp. 100-111.

[10] J. Dieudonne, "Sur le produit de composition," (II), *J. Math. Pures et Appl. 39* (1960), pp. 275-292.

[11] J. Dixmier, "Les moyennes invariantes dans les semi-groupes et leur applications," *Acta. Sci. Math.* (Szeged), *12* (1950), pp. 213-227.

[12] J. Dixmier, *Les algebres C^* et leur applications,* Gauthier-Villars, Paris, 1964.

[13] N. Dunford and J. Schwartz, "Linear Operators (part I). *Interscience* (1958), New York.

[14] W. Eberlein, "Abstract ergodic theorems and weak almost periodic functions," *Trans. Amer. Math. Soc. 67*

[15] W. Emerson and F. Greenleaf, "Covering properties and Følner conditions for locally compact groups," (to appear, *Math. Zeit.*).

[16] J. Feldman and F. P. Greenleaf, "Measurable transversals in locally compact groups," (*Pacific Math. J.*).

[17] J. M. G. Fell, "Weak containment and induced representations of groups," *Canadian Math. J. 14* (1962), pp. 237-268.

[18] ——, "The dual spaces of C^*-algebras," *Trans. Amer. Math. Soc. 94* (1960), pp. 365-403.

[19] E. Følner, "On groups with full Banach mean value," *Math. Scand. 3* (1955), pp. 243-254.

[20] H. Furstenberg, "A Poisson formula for semi-simple Lie groups," *Annals of Math. 77* (1963), pp. 335-386.

[21] I. Glicksberg, "On convex hulls of translates," *Pacific Math. J. 13* (1963), pp. 97-113.

[22] I. Glicksberg and K. de Leeuw, "The decomposition of certain group representations," *J. d'Analyse Math. 15* (1965), pp. 135-192.

[23] ——, "Applications of almost periodic compactifications," *Acta Math. 105* (1961), pp. 63-97.

[24] R. Godement, "Les functions de type positif et la theorie des groups," *Trans. Amer. Math. Soc. 63*(1948), pp. 1-84.

[25] E. Granirer, "On Baire measures on *D*-topological spaces," *Fund. Math., 60* (1967), pp. 1-22.

[26] ——, "On amenable semigroups with a finite dimensional set of invariant means (I + II)," *Ill. Math. J. 7* (1963), pp. 32-48 and 49-58.

[27] ——, "On the invariant means on topological semigroups and on topological groups," *Pacific Math. J. 15* (1965), pp. 107-140.

[28] F. Greenleaf, "Følner's condition for locally compact groups," (unpublished manuscript. 1967).

[29] ——, "Norm decreasing homomorphisms on group algebras," *Pacific Math. J. 15* (1965), pp. 1187-1219.

[30] A. Grothendieck, "Criteres de compacite dans les espaces functionnels genereaux," *Amer. J. Math. 74* (1952), pp. 168-186.

[31] P. Halmos, *Measure Theory,* Van Nostrand, New York, 1950.

[32] F. Hausdorff, *Grundzuge der Mengenlehre,* Leipzig (1914).

[33] S. Helgason, *Differential Geometry and Symmetric Spaces,* Academic Press, New York, 1962.

[34] E. Hewitt and K. Ross, *Abstract Harmonic Analysis, (v. 1).* Academic Press, New York, 1963.

[35] G. Hochschild, *The Structure of Lie Groups,* Holden-Day, San Francisco, 1965.

[36] A. Hulanicki, "Groups whose regular representation weakly contains all unitary representations, *Studia Math. 24* (1964), pp. 37-591

[37] ——, "Means and Følner conditions on locally compact groups," *Studia Math. 27* (1966), pp. 87-104.

[38] K. Iwasawa, "Some types of topological groups," *Annals of Math. 60* (1948), pp. 507-549.

[39] K. Jacobs, "Ergodentheorie und fastperiodische Funktionen auf Halbgruppen," *Math. Z. 64* (1956), pp. 298-338.

[40] ——, "Fastperiodizitatseigenschaften allgemeiner Halbgruppen in Banachraumen," *Math. Z. 67* (1957), pp. 83-92.

[41] S. Kakutani and K. Kodaira, "Uber das Haarsche mass in der lokal bikompakten gruppen," *Proc. Imp. Acad.* (Tokyo) *20* (1944), pp. 444-450.

[42] R. Kaufman, "Remark on invariant means," *Proc. Amer. Math. Soc. 18* (1967), pp. 120-122.

[43] Y. Kawada, "Uber den mittelwert der messbaren fastperiodischen funktionen auf einer gruppe," *Proc. Imp. Acad.* (Tokyo) *19* (1943), pp. 264-266.

[44] J. Kelley, I. Namioka, et al., *Linear Topological Spaces,* Van Nostrand, New York, 1963.

[45] H. Leptin, "On a certain invariant of a locally compact group," *Bull. Amer. Math. Soc. 72* (1966), pp. 870-874.

[46] ——, "On locally compact groups with invariant means," (to appear).

[47] ——, "Faltungen von Borelschen Massen mit L^p-Funktionen auf lokal kompakten Gruppen," *Math. Ann. 163* (1966), pp. 111-117.

[48] E. R. Lorch, "The integral representation of weakly almost periodic transformations in reflexive vector spaces," *Trans. Amer. Math. Soc. 49* (1941), pp. 18-40.

[49] E. R. Lorch, "A calculus of operators in reflexive vector spaces," *Trans. Amer. Math. Soc. 45* (1939), pp. 217-234.

[50] ——, "Bicontinuous linear transformations in certain vector spaces," *Bull. Amer. Math. Soc. 45* (1939), pp. 564-569.

[51] G. Mackey, "Induced representations of locally compact groups I," *Annals of Math. 55* (1952), pp. 101-139.

[52] T. Mitchell, "Constant functions and left invariant means on semigroups," *Trans. Amer. Math. Soc. 119* (1965), pp. 244-261.

[53] D. Montgomery and L. Zippin, "Topological transformation groups," *Interscience,* New York, 1955.

[54] I. Namioka, "Følner's condition for amenable semigroups," *Math. Scand. 15* (1964), pp. 18-28.

[55] ——, "On a recent theorem by H. Reiter," *Proc. Amer. Math. Soc. 17* (1966), pp. 1101-1102.

[56] I. Namioka and E. Asplund, "A geometric proof of Ryll-Nardzewski's fixed point theorem," (to appear).

[57] M. Rajagopalan, "L^p conjecture for locally compact groups (Parts I and II)," (to appear).

[58] H. Reiter, "The convex hull of translates of a function in L^1," *J. London Math. Soc. 35* (1960), pp. 5-16.

[59] ——, "Investigations in harmonic analysis," *Trans. Amer. Math. Soc. 73* (1952), pp. 401-427.

[60] ——, "Uber L^1-Raume auf gruppen, I and II," *Monatsh. Math. 58* (1954), pp. 73-76.

[61] ——, "Une propriete analytique d'une certaine classe de groups localement compacts," *C. Rend. Acad. Sci. 254* (1962), pp. 3627-3629.

[62] H. Reiter, "Sur les groupes de Lie semi-simples connexes," *C. Rend. Acad. Sci. 255* (1962), pp. 2883-2884.

[63] ——, "Sur la propriete (P_1) et les functions de type positif," *C. Rend. Acad. Sci. 258* (1964), pp. 5134-5135.

[64] ——, "On some properties of locally compact groups," *Indag. Math. 27* (1965), pp. 687-701.

[65] N. Rickert, "Some properties of locally compact groups," (to appear *Trans. Amer. Math. Soc.,* 1967).

[66] ——, "Amenable groups and groups with the fixed point property," (to appear).

[67] W. G. Rosen, "On invariant means over compact semigroups," *Proc. Amer. Math. Soc. 7* (1957). pp. 1076-1082.

[68] C. Ryll-Nardzewski, "On fixed points of semi-groups of endomorphisms of linear spaces," *Proceedings of the Fifth Berkeley Symposium on Math. Statistics and Probability, v. II,* Berkeley (1966).

[69] J. Stegeman, "On a property concerning locally compact groups," *Indag. Math. 27* (1965), pp. 702-703.

[70] B. Sz.-Nagy, "Uniformly bounded linear transformations in Hilbert space," *Acta Math.* (Szeged), *11* (1947), pp. 152-157.

[71] O. Takenouchi, "Sur une classe de fonctions de type positif sur une groupe localement compact," *Math. J. Okayama Univ. 4* (1955), pp. 143-173.

[72] J. von Neumann, "Zur allgemeinen theorie des masses," *Fund. Math. 13* (1929), pp. 73-116.

[73] H. Yoshizawa, "Some remarks on unitary representations of free groups," *Osaka Math. J. 9* (1951), 55-82.

[74] K. Yosida, *Functional Analysis,* Academic Press, New York, 1965.

[75] W. Zelazko, "On the algebras L_p of locally compact groups," *Coll. Math. 10* (1963), pp. 49-52.

[76] W. Emerson and F. P. Greenleaf, "Asymptotic behavior of products $E^p = E + \cdots + E$ in locally compact abelian groups," (to appear).

INDEX